THE MICROBIOLOGICAL RISK
ASSESSMENT OF FOOD

THE MICROBIOLOGICAL RISK ASSESSMENT OF FOOD

S.J. Forsythe
Department of Life Sciences
Nottingham Trent University

Blackwell
Science

© 2002 by Blackwell Science Ltd,
a Blackwell Publishing Company
Editorial Offices:
Osney Mead, Oxford OX2 0EL, UK
   Tel: +44 (0)1865 206206
Blackwell Science, Inc., 350, Main Street,
Malden, MA 02148-5018, USA
   Tel: +1 781 388 8250
Iowa State Press, a Blackwell Publishing
Company, 2121 State Avenue, Ames, Iowa
50014-8300, USA
   Tel: +1 515 292 0140
Blackwell Science Asia Pty, 54 University
Street, Carlton, Victoria 3053, Australia
   Tel: +61 (0)3 9347 0300
Blackwell Wissenschafts Verlag,
Kurfürstendamm 57, 10707 Berlin, Germany
   Tel: +49 (0)30 32 79 060

First published 2002 by Blackwell Science Ltd

Library of Congress
Cataloging-in-Publication Data
is available

ISBN 0-632-05952-4

A catalogue record for this title is available
from the British Library

Set in 10.5/12.5 Garamond
by DP Photosetting, Aylesbury, Bucks
Printed and bound in Great Britain by
MPG Books Ltd, Bodmin, Cornwall

For futher information on
Blackwell Science, visit our website:
www.blackwell-science.com

Cover picture: *Lactobacillus casei* Shirota by courtesy of Yakult UK Ltd.

CONTENTS

Preface		*vii*
1	**Food-borne Microbial Pathogens in World Trade**	**1**
	1.1 Food-borne microbial pathogens	1
	1.2 Chronic sequelae following food-borne illness	8
	1.3 Emergence and re-emergence of food-borne pathogens and toxins	10
	1.4 Changes in host susceptibility and exposure	12
	1.5 Risk of food poisoning	17
	1.6 The cost of food-borne diseases	18
	1.7 International control of microbiological hazards in foods	20
2	**Food Safety, Control and HACCP**	**34**
	2.1 Introduction	34
	2.2 HACCP adoption	35
	2.3 Outline of HACCP	37
	2.4 Control at source	45
	2.5 Product design and process control	45
	2.6 Microbial response to stress	53
	2.7 Predictive modelling	55
	2.8 Microbiological criteria	61
3	**Risk Analysis**	**66**
	3.1 Introduction	66
	3.2 Overview of microbiological risk assessment	69
	3.3 Risk assessment	75
	3.4 Risk management	103
	3.5 Food safety objectives	109
	3.6 Risk communication	110

4 Application of Microbiological Risk Assessment   113
  4.1 Introduction   113
  4.2 *Salmonella* spp.   113
  4.3 *Campylobacter jejuni* and *C. coli*   130
  4.4 *Listeria monocytogenes*   142
  4.5 Enterohaemorrhagic *E. coli* (EHEC); *E. coli* O157:H7   154
  4.6 *Bacillus cereus*   161
  4.7 *Vibrio parahaemolyticus*   165
  4.8 Mycotoxins   168
  4.9 Rotaviruses   173

**5 Future Developments in Microbiological
Risk Assessment   175**
  5.1 Introduction   175
  5.2 International methodology and guidelines   175
  5.3 Data   176
  5.4 Training courses and use of resources   177
  5.5 Microbiological risk assessment support system   180

Glossary of terms   181

References   185

Internet directory   200

Index   205

PREFACE

Why write a book about microbiological risk assessment? I confess that having embarked on this project I have subsequently asked myself that very question many, many times. The answer, at least in part, is that when I wrote *The Microbiology of Safe Food* in 2000 it was to emphasize that we eat food containing bacteria (even potentially pathogenic ones) daily, but at an 'acceptable' level. Also in the implementation of HACCP, one designs Critical Control Points which eliminate or reduce the hazard to an 'acceptable' level. However, the obvious question is, what is an acceptable level? But it is not an easy question to answer. For instance, are you a healthy middle aged person with no notable medical history? If so, then congratulations, but you're a dying breed. Nevertheless, your immune system is adept at dealing with the daily dose of micro-organisms. However, if you are very young or elderly then you are more susceptible. Hence governments and food manufacturers need to consider the consumer profile. This is where microbiological risk assessment is needed and hence is the answer to my question, what is an acceptable level?

In recent years there have been considerable achievements in microbiological detection methods, ranging from improved selective media, ATP bioluminescence and DNA probes. However, despite these many significant developments (some with which I'm very pleased to have been involved) their application for end-product testing or as a validation method for HACCP is flawed. This can be illustrated by combining a sampling plan with the operating characteristic curve which lead to 'consumer risk, producer risk' statistics. In a nutshell, a batch of food product can be 30% defective and if you test five samples with a sampling plan of $n = 5$, $c = 2$ then you'll accept that batch of food 85% of the time. Therefore we need a proactive approach to food safety, and that is encompassed in the HACCP approach and now in microbiological risk assessment.

Whilst preparing this book two others were published: one is by Bob Mitchell and is the straightforward *Practical Microbiological Risk*

Analysis and the other is by Phil Voysey (CCFRA, 2000) entitled *An Introduction to the Practice of Microbiological Risk Assessment for Food Industry Applications*. This left me to decide what angle I should emphasize, and after many abandoned versions here it is. Like Schlundt (Risk assessment advice, WHO online training course, see Internet Directory) I believe that microbiological risk assessment is the 'third wave' of food safety tools, the first being good hygienic practice and the second HACCP. Currently microbiological risk assessment is at the early stage of development and implementation, equivalent to HACCP in the late 1980s. Hence this book does not purely cover risk assessment but also related topics, such as predictive microbiology and sampling plans (which might equate in time to food safety objectives).

What next? Well four years ago in 1997 when I up-dated Pat Hayes' book *Food Hygiene, Microbiology and HACCP* I used a 286 PC. In 1998– 1999 *The Microbiology of Safe Food* was written on a 90 MHz Pentium, but with Internet links (see www.theagarplate.com). This current book, written during summer 2001, has used copious pdf files on the Web and a 330 MHz Pentium 4. In three years' time, we may be using ebooks, yet somehow I doubt if we'll have the Star Trek tricorder and hear a voice saying, 'Mr Spock, there's a salmonella over there'. Somehow technology progresses but the problems don't go away.

As usual, such an undertaking does not happen without the considerable support of family and friends. I must in particular thank Samantha Morris for her proofreading and express my kindest thoughts to Claire. My children, James and Rachel, have patiently tolerated my absence and my wife Debbie has been considerably long-suffering. To them all I owe a considerable debt of thanks.

Steve Forsythe
October 2001

1

FOOD-BORNE MICROBIAL PATHOGENS IN WORLD TRADE

1.1 Food-borne microbial pathogens

Microbial food poisoning is caused by a variety of micro-organisms with various incubation periods and duration of symptoms (Table 1.1). Organisms such as *Salmonella* and *Escherichia coli* O157:7 are well known by the general public. But there are also viruses and fungal toxins which are been relatively poorly studied and in the future we may more fully recognise their contribution to the general incidence of food

Table 1.1 Common food poisoning micro-organisms.

Micro-organism	Incubation period	Duration of illness
Aeromonas species	Unknown	1–7 days
C. jejuni	3–5 days	2–10 days
E. coli		
ETEC	16–72 hours	3–5 days
EPEC	16–48 hours	2–7 days
EIEC	16–48 hours	2–7 days
EHEC	72–120 hours	2–12 days
Hepatitis A	3–60 days	2–4 weeks
L. monocytogenes	3–70 days	Variable
Norwalk-like virus	24–48 hours	1–2 days
Rotavirus	24–72 hours	4–6 days
Salmonellae	16–72 hours	2–7 days
Shigellae	16–72 hours	2–7 days
Y. enterocolitica	3–7 days	1–3 weeks

poisoning. Micro-organisms causing food poisoning are found in a diverse range of foods (Table 1.2). They have a wide range of virulence factors, and may elicit a wide spectrum of adverse responses that may be acute, chronic or intermittent. Some bacterial pathogens, such as salmonellae, are invasive and may cause bacteraemia and generalised infections. Other pathogens produce toxins that cause severe damage in susceptible organs such as the kidney (for example *E. coli* O157:H7). Complications may also arise by immune-mediated reactions (e.g. reactive arthritis and Guillain-Barré syndrome) where the immune response to the pathogens is also directed against the host tissues. The complications from enteritis normally require medical care and frequently result in hospitalisation. There may be a substantial risk of mortality in relation to sequelae, and not all patients may recover fully but may suffer from residual symptoms which may last for a lifetime. Therefore, despite the low probability of these complications, the public health burden may be considerable (see Section 1.2). Obviously there is a greater risk of mortality in the elderly and severely immunocompromised following an acute infection.

Because consumers are unaware that there is a potential problem with the food, a significant amount of contaminated food is ingested and hence they become ill. Consequently, it is hard to trace which food was the original cause of food poisoning because the consumers will not recall noticing anything appropriate in their recent meals. They are, however, likely to recall food which smelt 'off' or looked 'discoloured'; however, these changes are related to food spoilage and not food poisoning.

Table 1.2   Sources of food-borne pathogens.

Food	Pathogen	Incidence (%)
Meat, poultry and eggs	*C. jejuni*	Raw chicken and turkey (45–64)
	Salmonella spp.	Raw poultry (40–100), pork (3–20), eggs (0.1) and shellfish (16)
	St. aureus[a]	Raw chicken (73), pork (13–33) and beef (16)
	Cl. perfringens[b] *Cl. botulinum*	Raw pork and chicken (39–45)
	E. coli O157:H7	Raw beef, pork and poultry
	B. cereus[b]	Raw ground beef (43–63), cooked meat (22)
	L. monocytogenes	Red meat (75), ground beef (95)
	Y. enterocolitica Hepatitis A virus *Trichinella spiralis* Tapeworms	Raw pork (48–49)

(Contd)

Table 1.2   *(Contd)*

Food	Pathogen	Incidence (%)
Fruit and vegetables	*C. jejuni*	Mushrooms (2)
	Salmonella spp.	Artichoke (12), cabbage (17), fennel (72), spinach (5)
	St. aureus[a]	Lettuce (14), parsley (8), radish (37)
	L. monocytogenes	Potatoes (27), radishes (37), bean sprouts (85), cabbage (2), cucumber (80)
	Shigella spp.	
	E. coli O157:H7	Celery (18) and coriander (20)
	Y. enterocolitica	Vegetables (46)
	A. hydrophila	Broccoli (31)
	Hepatitis A virus	
	Norwalk-like virus	
	Giardia lamblia	
	Cryptosporidium spp.	
	Cl. botulinum	
	B. cereus[b]	
	Mycotoxins	
Milk and dairy products	*Salmonella* spp.	
	Y. enterocolitica	Milk (48–49)
	L. monocytogenes	Soft cheese and patc (4–5)
	E. coli	
	C. jejuni	
	Shigella spp.	
	Hepatitis A virus	
	Norwalk-like virus	
	St. aureus[a]	
	Cl. perfringens[b]	
	B. cereus[b]	Pasteurised milk (2–35), milk powder (15–75), cream (5–11), ice cream (20–35)
	Mycotoxins	
Shellfish and fin fish	*Salmonella* spp.	
	Vibrio spp.	Raw seafood (33–46)
	Shigella spp.	
	Y. enterocolitica	
	B. cereus	Fish products (4–9)
	E. coli	
	Cl. botulinum[b]	
	Hepatitis A virus	
	Norwalk-like virus	
	Giardia lamblia	
	Cryptosporidium spp.	
	Metabolic by-products	
	Algal toxins	

(Contd)

Table 1.2   *(Contd)*

Food	Pathogen	Incidence (%)
Cereals, grains, legumes and nuts	*Salmonella* spp. *L. monocytogenes* *Shigella* spp. *E. coli* *St. aureus*[a] *Cl. botulinum*[b] *B. cereus*[b] Mycotoxins[a]	Raw barley (62–100), boiled rice (10–93), fried rice (12–86)
Spices	*Salmonella* spp. *St. aureus*[a] *Cl. perfringens*[b] *Cl. botulinum*[b] *B. cereus*[b]	Herbs and spices (10–75)
Water	*Giardia lamblia*	Water (30)

[a] Toxin not destroyed by pasteurisation.
[b] Spore-forming organism; not killed by pasteurisation.

Sources: Various including Synder (Hospitality Institute of Technology and Management, Web address in the Internet Directory) and ICMSF (1998a).

Food poisoning micro-organisms are normally divided into two groups:

- Infections: *Salmonella* serotypes, *Campylobacter jejuni* and pathogenic *Escherichia coli*.
- Intoxications: *Bacillus cereus, Staphylococcus aureus, Clostridium botulinum*.

The first group are micro-organisms which multiply in the human intestinal tract, whereas the second group produce toxins either in the food or during passage in the intestinal tract. This division is very useful to help recognise the routes of food poisoning. Vegetative organisms are killed by heat treatment, whereas spores (produce by *B. cereus* and *Clostridium perfringens*) may survive and hence germinate if the food is not kept sufficiently hot or cold.

An alternative grouping would be according to severity of illness. This approach is useful in setting microbiological criteria (sampling plans) and risk analysis (Notermans *et al.* 1995). The International Commission on Microbiological Specifications for Foods (ICMSF 1986; currently under revision) divided the common food-borne pathogens into such groups to aid decision making of sampling plans (Section 2.8.1). The ICMSF groupings are given in Table 1.3.

Table 1.3   Microbiological hazards categorisation according to ICMSF.

Effects of hazards	Pathogen
Categorisation of common foodborne pathogens (ICMSF 1986)	
(1) Moderate, direct, limited spread, death rarely occurs	*B. cereus, C. jejuni, Cl. perfringens, St. aureus, Y. enterocolitica, Taenia saginata, Toxoplasma gondii*
(2) Moderate, direct, potentially extensive spread, death or serious sequelae can occur. Considered severe	Pathogenic *E. coli, S.* Enteritidis and other salmonellae other than *S.* Typhi and *S.* Paratyphi, shigellae other than *Sh. dysenteriae, L. monocytogenes*
(3) Severe, direct	*Cl. botulinum* types A, B, E and F, hepatitis A virus, *Sh. dysenteriae, S.* Typhi and *S.* Paratyphi A, B and C, *T. spiralis*
Proposed up-dated categorisation (ICMSF unpublished)	
(1) Food poisoning organisms causing moderate, not life threatening, no sequelae, normally short duration, self-limiting	*B. cereus* (including emetic toxin), *Cl. perfringens* type Λ, Norwalk-like viruses, *E. coli* (EPEC, ETEC), *St. aureus, V. cholerae* non-O1 and non-O139, *V. parahaemolyticus*
(2) Serious hazard, incapacitating but not life-threatening, sequelae rare, moderate duration	*C. jejuni, E. coli, S.* Enteritidis, *S.* Typhimurium, shigellae, hepatitis A, *L.* Monocytogenes, *Cryptosporidium parvum*, pathogenic *Y. enterocolitica, Cyclospora cayetanensis*
(3) Severe hazard for general population, life-threatening, chronic sequelae, long duration	Brucellosis, botulism, EHEC (HUS), *S.* Typhi, *S.* Paratyphi, tuberculosis, *Sh. dysenteriae*, aflatoxins, *V. cholerae* O1 and O139.
(4) Severe hazard for restricted populations, life-threatening, chronic sequelae, long duration	*C. jejuni* O19 (GBS), *Cl. perfringens* type C, hepatitis A, *Cryptosporidium parvum, V. vulnificus, L. monocytogenes*, EPEC (infant mortality), infant botulism, *Ent. sakazakii*

Detailed descriptions of certain food-borne pathogens are given in Chapter 4 together with related risk assessments. Further information for other micro-organisms can be found in *The Microbiology of Safe Food* (Forsythe 2000) and extensive details can be found in the numerous ICMSF publications (ICMSF 1986, 1988, 1996a, 1997, 1998a,c).

Despite an increasing awareness and understanding of food- and water-borne micro-organisms, these diseases remain a significant problem and are an important cause of reduced economic productivity. While everyone is susceptible to food-borne diseases, there are a growing number of people who are more likely to experience such diseases, often with more severe consequences. These people include infants and young children, pregnant women, those who are immunocompromised and the elderly. There is evidence that the microbial causes of gastroenteritis vary with age and that viral agents are probably the major infectious agent in children under 4 years (Fig. 1.1). There is also a difference between the sexes (Fig. 1.2) which is possibly due to differences in personal hygiene, i.e. males have a lower tendency to wash their hands after going to the toilet.

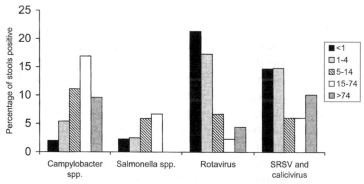

Fig. 1.1   Variation in causative agent of gastroenteritis with age (Forsythe 2000).

Food-borne diseases are a major contributor to the estimated 1.5 billion annual episodes of diarrhoea in children under the age of 5 years. Children in developing countries suffer two or three episodes of diarrhoea per year, and in some cases as many as ten episodes. Up to 70% of such episodes in children under 5 years old have been attributed to contaminated food. Weaning foods contaminated with pathogenic strains of *E. coli* are considered to be the cause of 25–30% of diarrhoeal disease episodes in developing countries. A serious consequence of diarrhoeal disease is the effect on the nutritional status and immune systems of infants and children. Repeated episodes lead to a reduction in food intake, aggravated by loss of nutrients due to malabsorption and vomiting, fever and impaired resistance to other infections (often respiratory); hence the child becomes caught up in a vicious cycle of malnutrition and infection. Many do not survive under these circumstances, and some 13 million children under 5 years old die annually in this way. By the year 2025 more than 1000 million

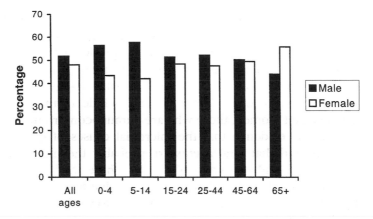

Fig. 1.2 Variation in gastroenteritis incidence with sex (Forsythe 2000).

of the world's population will be aged over 60 years. More than two-thirds of them will live in developing countries. Growing old means increased risk from food-borne diseases. It is not surprising then, that in some countries one person in four is especially at risk of food-borne disease. A wide range of food-borne diseases are prevalent in developing countries. These include cholera, salmonellosis, campylobacteriosis, shigellosis, typhoid, poliomyelitis, brucellosis, amoebiasis and *E. coli* infections.

The exact number of annual 'food poisoning' cases can only be estimated. In many instances, only a small proportion of cases seek medical help and not all are investigated. In the past it has been assumed that in industrialised countries less than 10% of the cases were reported, while in developing countries reported cases probably account for less than 1% of the total. However, a better estimate is being achieved using sentinel studies as reported for the USA (Mead *et al.* 1999), UK (Wheller *et al.* 1999) and The Netherlands (de Wit *et al.* 2001). In the USA, it has been estimated that 76 million cases of food-borne diseases may occur each year, resulting in 325 000 hospitalisations and 5000 deaths (Mead *et al.* 1999). The UK study similarly estimates that the proportion of the public experiencing gastroenteritis due to food-borne pathogens is 20% each year and perhaps up to 20 people per million die. The more recent sentinel study in The Netherlands estimated the number of microbial food-borne illnesses to be 79.7 per 10 000 person years (de Wit *et al.* 2001). Notermans and van der Giessen (1993) estimated the number to be 30% of the population per year.

Because food poisoning symptoms are generally mild and only last a few days, the affected person usually recovers without seeking medical care. Nevertheless, those at higher risk, such as the very young, pregnant and elderly, may suffer more serious, debilitating and even life-threatening illness. These cases have previously been largely overlooked in determinations of the human burden of food poisoning. The following section is aimed at re-addressing the balance.

1.2   Chronic sequelae following food-borne illness

The potential for chronic sequelae (secondary complications) has been recently recognised together with the variability of the human response (Table 1.4). It has been estimated that chronic sequelae occur in 2–3% of food-borne cases and may last in terms of weeks or even months. These

Table 1.4   Chronic sequelae following food-borne infection (Mossel 1988).

Disease	Associated complication
Brucellosis	Aotitis, orchitis, meningitis, pericarditis, spondylitis
Campylobacteriosis	Arthritis, carditis, cholecystitis, colitis, endocarditis, erythema nodosum, Guillain–Barré syndrome, haemolytic uraemic syndrome, meningitis, pancreatitis, septicaemia
E. coli (EPEC & EHEC types) infections	Erythema nodosum, haemolytic uraemic syndrome, seronegative arthropathy
Listeriosis	Meningitis, endocarditis, osteomyelitis, abortion and stillbirth, death
Salmonellosis	Aortitis, cholecystitis, colitis, endocarditis, orchitis, meningitis, myocarditis, osteomyelitis, pancreatitis, Reiter's syndrome, rheumatoid syndromes, septicaemia, splenic abscess, thyroiditis
Shigellosis	Erythema nodosum, haemolytic uraemic syndrome, peripheral neuropathy, pneumonia, Reiter's syndrome, septicaemia, splenic abscess, synovitis
Taeniasis	Arthritis
Toxoplasmosis	Foetus malformation, congenital blindness
Yersiniosis	Arthritis, cholangitis, erythema nodosum, liver and splenic abcesses, lymphadenitis, pneumonia, pyomyositis, Reiter's syndrome, septicaemia, spondylitis, Still's disease

sequelae may be more serious than the original illness and result in serious long term disability or even death (Archer & Young 1988; Lindsay 1997; Bunning *et al.* 1997). The evidence that micro-organisms are implemented in chronic sequelae is not always certain because the chronic complications are unlikely to be epidemiologically linked to a food-borne illness.

A major symptom of food poisoning is diarrhoea which may consequently lead to anorexia and malabsorption. Severe cases of diarrhoea may last for months or years and can be caused by *Campylobacter jejuni*, *Citrobacter*, *Enterobacter* or *Klebsiella* enteric infections. Subsequently, the intestinal wall permeability may be altered so that significant quantities of unwanted proteins are absorbed which may induce atrophy.

Food-borne pathogens may interact with the immune system of the host to elude or alter the immunological process, which may subsequently induce a chronic disease. These interactions include the following:

- Antigenic heterogeneity (phase variation) whereby some micro-organisms have the ability to change their antigenic profile
- Sequestration, either intracellularly or in specific host sites
- Molecular mimicry, through either imitation (cross reaction) or adsorption of host protein
- Direct immune stimulation and/or suppression.

In addition, genetic susceptibility of the host may predispose humans to some types of infection. The following are of particular importance:

(1) Reactive arthritis and Reiter's syndrome (*Salmonella* serotypes, Section 4.2)
(2) Guillain–Barré syndrome (*C. jejuni*, Section 4.3)
(3) Haemolytic uraemic syndrome (*E. coli* O157, Section 4.5).

Less well studied chronic sequelae are as follows:

(1) Chronic gastritis due to *Helicobacter pylori*
(2) Crohn's disease and ulcerative colitis, possibly caused by *Mycobacterium paratuberculosis*
(3) Long term gastrointestinal and nutritional disturbances following infection by *C. jejuni*, *Citrobacter*, *Enterobacter* or *Klebsiella*
(4) Haemolytic anaemia due to *Campylobacter* and *Yersinia*
(5) Heart and vascular diseases caused by *E. coli*
(6) Atherosclerosis following *Salmonella* Typhimurium infection
(7) Personality changes following toxoplasmosis
(8) Graves disease (autoimmune disease) mediated by autoantibodies

to the thyrotropin receptor after *Yersinia enterocolitica* serotype O:3 infection

(9)   Severe hypothyroidism due to *Giardia lamblia* infection
(10) Crohn's disease, possibly due to *Mycobacterium paratuberculosis* (causative agent of Johne's disease in ruminants) via pasteurised milk. Other causative bacteria may be *Listeria monocytogenes*, *E. coli* and *Streptococcus* spp.
(11) Viral induction of autoimmune disorders, such as hepatitis A virus infection causing acute hepatitis with jaundice in adults. This is probably due to molecular mimicry
(12) Mycotoxins have a range of acute, subacute and chronic toxicities, with some molecules being carcinogenic, mutagenic and terato-genic (Section 4.8).

1.3   Emergence and re-emergence of food-borne pathogens and toxins

For various reasons, the number of identified food-borne pathogens has increased in recent years (Tauxe 1997). Emerging (and re-emerging) infections have been defined by the US National Research Council as 'new, recurring, or drug-resistant infections whose incidence in humans has increased in the last decades or whose incidence threatens to increase in the near future' (NRC 1993). Current surveillance methods only detect pathogens in 40–60% of patients suffering from gastroenteritis (de Wit *et al.* 2001). Hence there is considerable scope for new pathogens to be discovered or 'emerge'. Some emerging food-borne diseases are well characterised, but are considered as 'emerging' because the reporting of them has recently (over the past 10–15 years) become more common. Table 1.5 lists the various emerging food- and water-borne pathogens and toxins.

The (re-)emergence of certain food-borne pathogens and toxins is due to a number of causes (Lindsay 1997).

(1)   Weakened or collapsed public health infrastructure for epidemic disease control due to economic problems, changing health policies, civil strife and war
(2)   Poverty, uncontrolled urbanisation and population displacements
(3)   Environmental degradation, and contamination of water and food sources
(4)   Ineffective infectious disease control programmes
(5)   Newly appeared organisms in the microbial population, such as those resulting from inappropriate use of antibiotics, including

Table 1.5 Emerging pathogens and toxins in food

Bacteria	*E. coli* O157:H7, enteroaggregative *E. coli* (EAEC), *V. cholerae*, *V. vulnificus*, *St. parasanguinis*, *Mycobacterium paratuberculosis*, *L. monocytogenes*, *S.* Typhimurium DT104, *S.* Enteritidis, *C. jejuni*, *Arcobacter* spp., *Enterobacter sakazakii*
Viruses	Hepatitis E, Norwalk virus and Norwalk-like virus
Protozoa	*Cyclospora cayetanensis, Toxoplasma gondii, Cryptosporidium parvum*
Helminths	*Anisakiasis simplexi* and *Pseudoterranova decipiens*
Prions	Bovine spongiform encephalitis, variant CJD
Mycotoxins	Fumonisins, zearalenone, trichothecenes, ochratoxins

     antibiotics used in animal production which are responsible for the rise of resistance to antimicrobial drugs
(6) Diseases crossing from animal to human populations with increasing frequency, especially when humans exploit new ecological zones, and intensification of animal food production (including that of fish) and industrialisation of food processing and distribution become global practices
(7) Increased potential for spread of disease through globalisation of travel and trade, including that of processed and raw foodstuffs of vegetal and animal origin
(8) Dispersal by new vehicles of transmission
(9) Causes recently identified as the result of increased knowledge or new methods of identification, though they were previously widespread.

The factors that can affect the epidemiology of emerging food pathogens are given in Table 1.6.

    Micro-organisms have evolved many adaptation mechanisms to survive and persist in otherwise 'unfavourable' growth conditions (Lederberg 1997). They can exchange genetic material (e.g. by conjugation, transduction and transformation) and hence acquire new gene sequences. Horizontal gene transfer is now recognised as one route by which toxin genes can be distributed to new bacterial strains (e.g. origin of *E. coli* O157:H7). Gene sequences called 'pathogenicity islands' which encode for specific virulence factors can be recognised by their percentage GC content which differs from the rest of the bacterial genome. Likewise, though less well understood, viruses may interact with host genomes and

Table 1.6   Factors affecting the epidemiology of emerging food pathogens.

Microbial adaptation through natural selection; antibiotic usage can select for antibiotic resistant strains, e.g. *S.* Typhimurium DT104
New foods and food preparation technologies, e.g. BSE and vCJD
Changes in host susceptibility, e.g. increase in population age
Changes in lifestyle such as an increased consumption of 'convenience food' and subsequent risk to *L. monocytogenes*
Increasing international trade and travel enabling the rapid spread of pathogens worldwide, e.g. *E. coli* O157:H7
New food vehicles of transmission being recognised, e.g. *Mycobacterium paratuberculosis* (plausible)

emerge in new susceptible populations and vehicles, with less than predictable outcomes. Food-borne pathogens often have an animal reservoir from which they can spread to humans, although they frequently do not cause illness in the primary host. Because of the considerable increase in international travel and trade, food-borne pathogens can rapidly be spread world-wide.

There has been considerable concern over the emergence of antibiotic resistant food-borne pathogens such as *S.* Typhimurium DT104 and *C. jejuni*. One of the most publicised concerns is the use of fluoroquinolone antibiotics in both veterinary and medical practice. The risk assessment for fluoroquinolone resistance in *C. jejuni* is covered in Section 4.3.

1.4   Changes in host susceptibility and exposure

The days of locally produced food being processed, distributed and consumed in the same locality have significantly decreased in recent decades. The regional, national and global food chain has required parallel changes in food science and technology, including preservation. At the same time, there have been social changes such as an increasing number of meals being consumed outside the home environment and also an ageing population. Public exposure to a food-borne pathogen may change due to changes in the processing (e.g. exposure to bovine spongiform encephalopathy (BSE)), changes in consumption patterns and the globalisation of the food supply chain. Many 'risk factors' influence host susceptibilty to infection (Section 3.3.4). These may be:

- Pathogen-related: ingested dose, virulence
- Host-related: age, immune status, personal hygiene, genetic susceptibility

- Diet-related: nutritional deficiencies, ingestion of fatty or highly buffered foods.

1.4.1 Processing methods and globalisation of the food industry

The globalisation of the food supply is recognised as a major trend contributing to food safety problems. Pathogenic micro-organisms are not contained within a single country's borders. Additionally, tourism and increased cultural interests may lead to new eating habits, such as the consumption of 'sushi' (origin in Japan) in Western countries. The continuous increase in international trade has been achievable partly through advances in food manufacturing and processing technologies together with improvements in transportation. Regional trade arrangements and the overall impact of the Uruguay Round Agreements (Section 1.7.1) have reduced many tariff- and subsidy-related constraints to free trade, encouraging increased production and export from the countries with the most cost-effective production means. However, many exporting countries do not have the infrastructure to ensure high levels of hygienic food manufacture.

The continuing integration and consolidation of agriculture and food industries and the globalisation of food trade are changing the patterns of food production and distribution, as well as supply and demand. Production of raw materials is increasingly concentrated in fewer, specialised and larger production units. These changes may have many benefits, but also present new challenges to safe food production, and may have widespread repercussions on health. The pressure to produce food for export is very significant in developing economies and can lead to improper agricultural practices. The consequences may include the following:

- Accidental or sporadic low level microbial contamination of a single product, which can result in a major epidemic of food-borne illness
- High levels of mycotoxins, often resulting from poor storage and handling conditions
- High pesticide residues in food
- Industrial contamination of food with metals and chemicals such as polychlorinated biphenyls (PCBs) and dioxins.

Changes in farming practices, such as the introduction of new plant varieties and new crop rotation practices, can introduce or increase the presence of hazards such as mycotoxin contamination of the crop. Because of the extensive distribution of food from a single source, the potential for many food consumers to be affected by a localised contamination has increased. This was demonstrated by the dioxin incident in Belgium and the BSE outbreak in England, where a risk generated in one country had global implications.

The consumer demand for food which is less processed and contains fewer additives means that food processors have less choice in their preservation methods (Zink 1997). However, because the consumer also expects a long shelf-life, this can lead to problems with psychrotrophic pathogenic bacteria, such as *Listeria*, *Yersinia* and *Aeromonas* (in sous-vide foods, for example). Hence food processors are investigating new processing and preservation techniques, including high pressure treatment, ohmic heating and exposure to a pulsed electric field.

1.4.2   Changes in consumption pattern

Eating away from home is a major trend of recent years. Many of the meals eaten away from home require extensive food handling and/or are cold foods that are not cooked before consumption. Subsequently this leads to an increase in the number of people handling food and hence increases the potential for transmission of food-borne diseases from food handlers to consumers. It is plausible that because individuals spends less time in daily home food preparation, they have less knowledge regarding safe food preparation. Several studies have documented an increasing lack of knowledge about safe home food preparation or preservation practices, such as personal hygiene, the use of clean utensils and storage of food at the correct temperature.

1.4.3   Immunological status

The immune system may be partially compromised or undeveloped in new-born babies, the very young, pregnant women, those on medication or having an underlying illness, e.g. AIDS, and the elderly. Young children are more likely than adults to develop illnesses from selected pathogens (Fig. 1.1). Socioeconomic factors affect vulnerability. For example, in developed countries the case-fatality rate for typhoid is higher among individuals 55 years or older. In contrast, in developing countries, the higher risks of complication and death are for children from birth to 1 year of age and adults over 31 years (Gerba *et al*. 1996).

During pregnancy, women have a suppressed immune system to reduce the rejection of the foetus; consequently pregnant women have a greater risk of food poisoning due to *L. monocytogenes* than the general population. Infection of the foetus or the new-born babies from infected mothers can be extremely severe, resulting in abortion, stillbirth, or a critically ill baby that may present with an early onset (mortality rate

15–50%) or late onset (mortality rate 10–20%) form of listeriosis (Farber & Peterkin 1991; Farber *et al.* 1996). Neonates are uniquely susceptible to enterovirus infections such as those of coxsackie B and echovirus (Gerba *et al.* 1996).

Infections in the immunocompromised (non-pregnant) host constitute a new and severe aspect of the food safety problem. Advances in medical treatment have resulted in an increased number of immunosuppressed patients (e.g. cancer cases, organ transplant cases), and patients with serious underlying chronic diseases who may be at increased risk from food-borne infection and/or develop more severe illness. Obviously, enteric pathogens can easily cause persistent and generalised infection in the immunocompromised host. The majority of AIDS patients (50–90%) suffer from chronic diarrhoeal diseases which can be fatal (Morris & Potter 1997).

The proportion of elderly people in the population is increasing. It has been projected that in the USA one-fifth of the population will be over the age of 65 years by the year 2030. The increased susceptibility of the elderly may be due to a number of physiological factors, such as the senescence of the gut-associated lymphoid tissue and/or the decrease in gastric acid secretion, which reduce the natural barriers to gastrointestinal pathogens. Also, the immune system of the elderly is often weakened as a result of chronic illnesses. The incidence of salmonellosis, *Campylobacter* enteritis or *E. coli* O157:H7 infection appears to be higher in the elderly. Epidemiological studies have shown that the elderly experience higher case-fatality rates than the general population; for example, 3.8% versus 0.1% for *Salmonella*, 11.8% versus 0.2% for *E. coli* O157:H7 and 1% versus 0.01% for rotavirus (Gerba *et al.* 1996).

1.4.4 Malnutrition

On a global scale, the leading cause of increased host susceptibility to food-borne infections is probably malnutrition. In developing countries malnutrition affects about 800 million people. The region with the largest absolute number affected is Asia (524 million) and the region with the largest proportion of the population affected is Africa (28%): in some individual countries, the proportion can be as high as 30%. Malnutrition increases host susceptibility to food-borne infections through a number of mechanisms. It weakens epithelial integrity and may have a profound influence on cell-mediated immunity, with functional deficiencies in immunoglobulin and deficient phagocytosis. Malnutrition results in a 30-fold increase in the risk of diarrhoea-associated death (Morris & Potter 1997).

1.4.5   Changing perceptions and values

Public concerns on food related issues vary between countries and change with time (Sparks & Shepherd 1994). Before 1990, food additives were the focus of attention. In the 1990s the public became more aware of food-borne pathogens such as *Salmonella*, and in the late 1990s the cause for concern has been the link between 'mad cow disease' and variant CJD, and also (particularly in Europe) food biotechnology. Hence, members of the public nowadays have a greater awareness of 'food poisoning' and have become critical and cautious of new production methods. The adverse public attention to the food industry has led to distrust. This caution has been fed by numerous well publicised instances of food poisoning or 'food scares' which have cost the food industry considerable sums of money; see Table 1.7.

Table 1.7   Major food scares around the world.

Salmonella in beef, Aberdeen 1964
Salmonella in tinned salmon, Birmingham, UK 1986
Dioxins in poultry, Belgium 1999
Toxic mustard seed oil, India 1998
Chernobyl accident 1986, contamination of wide areas of Western Europe and subsequently food with radionuclides
Toxic cooking oil, Spain 1981
E. coli O157:H7, Japan
Listeria monocytogenes, France
E. coli O157:H7, Washington (others)
BSE-vCJD, UK
Genetically modified crops: despite scientific assurances the general public in certain countries, notably Europe, are unwilling to consume foods containing GM products

Though the food industry needs to produce food that is safe to eat, it must also manufacture food that is of the quality expected by the consumer. However, 'safe food' is an arbitrary term, meaning different things to different people, and does not necessarily equate with 'zero risk'. Hence risk management and risk communication are important to the food industry and are defined in Section 3.2. The 'zero risk' approach is not feasible; for example, the control of a food-borne pathogen requires the use of preservatives (with perceived toxicological risks) or heat treatment (possible production of carcinogens). Subsequently, the concept of 'threshold' means that there exists a limit under which the risk is inexistant or negligible. In medical (and consequently food) microbiology, the 'minimal infectious dose' has been the threshold. However, this approach has been re-evaluated in microbiological risk assessment to estimate the probability of infection (P_i) by a single cell (Section 3.3.5; Vose 1998).

1.5   Risk of food poisoning

As stated above, although industry and national regulators strive for production and processing systems which ensure that all food is 'safe and wholesome', a complete freedom from risks is an unattainable goal. Safety and wholesomeness are related to a level of risk that society regards as reasonable in the context, and in comparison with other risks in everyday life. Putting food poisoning into context is not an easy task due the high level of publicity which it receives in some countries. Table 1.8 summarises the risk of death in the next 12 months. One can see that death due to food poisoning has an equivalent risk to being a pedestrian who is struck by a car. However, such tragic incidences occur to individuals and only reach the local newspaper headlines, whereas food poisoning outbreaks involving a large number of people are more 'significant' to the media.

The reported increase in the number of 'food poisoning cases' is commonly cited by the media. However, these trends must be regarded with caution because they are the number of gastroenteritis cases which

Table 1.8   Risk of death during the next 12 months.

Cause	Mortality, chance of 1 in
Smoking 10 cigarettes a day	200
Natural causes, middle-aged	850
Death through influenza	5 000
Dying in a road accident	8 000
Flying	20 000
Pedestrian struck by car	24 000
Food poisoning	25 000
Dying in a domestic accident	26 000
Being murdered	100 000
Death in a railway accident	500 000
Electrocution	200 000
Struck by lightning	10 000 000
Eating beef on the bone	1 000 000 000

have been investigated and the causative organism identified. Not all cases have been shown to be due to food vectors, and increase in public awareness can increase the number of general public seeking medical help; in addition, improved detection methods could 'increase' the number of cases over time. In fact, across Europe and the United States the number of reported gastroenteritis cases has been decreasing (Anon. 1999a). Possibly due to changes in eating habits, there is a marked seasonality in food poisoning incidences with pathogens such as *Salmonella* and *Campylobacter*. As given in Section 1.1, sentinel studies indicate that the true incidence of gastroenteritis may be about 20%, though the mild nature of the illness means it passes unnoticed and is hence unreported.

1.6   The cost of food-borne diseases

In addition to human suffering, food-borne diseases can also be costly. Buzby and Roberts (1997a) estimated that in the USA the medical costs and productivity losses are in the range of US$6.6–37.1 billion (American billion). The cost of human illness due to only six bacterial pathogens is US$9.3–12.9 billion annually. Of these costs, US$2.9–6.7 billion are attributable to the food-borne bacteria *Salmonella* serovars, *C. jejuni, E. coli* O157:H7, *L. monocytogenes, St. aureus* and *Cl. perfringens*. Frenzen *et al.* (1999) estimated the cost of food-borne salmonellosis to be as high as US$2.3 billion annually for medical costs and lost productivity in the USA. Roberts (1996) estimated that the medical costs and value of lives lost due to just five food-borne infections in England and Wales was £300–700 million per year. In Australia, the cost of an estimated 11 500 food poisoning cases per day is A$2.6 billion per year (ANZFA 1999). Less recent estimations for Canada are US$1.3 billion loss due to food-borne pathogens (Todd 1989).

The Communicable Disease Center (USA) estimates that 95% of *Salmonella* infections are food-borne in origin. Consequently the FoodNet estimate of 1.4 million annual salmonellosis cases means that 1.3 million cases were due to consumption of foods contaminated by *Salmonella* (Frenzen *et al.* 1999). The estimated medical costs of *Salmonella* infections were based on the average medical care per case or each severity category, the estimated number of cases, and the 1998 average cost in the USA for each type of medical care. The value of a life ranges from US$8.3 and US$8.5 million at birth to US$1.4 and US$1.6 million at age 85 years and above, for males and females, respectively. Because approximately two-thirds of those dying from *Salmonella* infections were aged 65 years or more, the average foregone earnings per premature death were US$4.1 million for males and US$3.5 million for females. Medical care and lost productivity were calculated to be US$0.5 billion. Time lost from work equated US$0.9–12.8 billion. The annual cost of campylobacteriosis in the USA is approximately US$0.8–5.6 billion (Buzby & Roberts 1997b), and estimated total costs of campylobacter-associated Guillaine–Barré syndrome (GBS, an autoimmune reaction) are US$0.2–1.8 billion (Section 4.3). Hence reducing the prevalence of *Campylobacter* in food could save up to US$5.6 billion annually. The Swedish *Salmonella*-free poultry programme costs about US$8 million per year, but saves an estimated US$28 million per year in medical costs (Altekruse *et al.* 1993).

If the incidence of food-borne disease increases, the incidence of chronic sequelae, such as GBS may also rise. Chronic sequelae may occur in 2–3% of food-borne disease cases; hence the long-term consequences to the economy may be more detrimental than the acute disease. Thus there is a considerable need, even in the developed countries, for more systematic and aggressive steps to be taken to reduce significantly the risks posed by microbiological food-borne diseases. Since the cost of 'gastroenteritis' in developed countries is determined to be billions of US dollars, how much more is the human cost of food- and water-borne disease in developing countries? The World Health Organisation (WHO) estimates that, world-wide, almost 2 million children die every year from diarrhoea. A significant proportion of deaths are caused by microbiologically contaminated food and water. Hence this is a *major* challenge for the future, which requires a global response. Therefore this book frequently refers to the activities of international organisations such as the Codex Alimentarius Commission (CAC) and the WHO.

The impact of food losses due to microbial contamination is also considerable. World-wide losses of grain and legumes are estimated to be at least 10% of production, and for non-grain staples, vegetables and fruits, the loss could be as high as 50%. Food contamination affects trade in two ways. Firstly, contaminated food may be rejected if the level of contaminants is above the limits permitted by importing countries. For example, during a 3 month period from January to March 1980, food imports valued at about US$20 million were rejected in the USA on account of food contamination with moulds and aflatoxins. Secondly, a country's reputation in food safety may cause a decrease in trade as well as in tourism. The epidemic of cholera in Peru cost the country over US$700 million in 1991 due to the loss of food exports and tourism, in addition to medical care costs. The socioeconomic impact of unrelated issues such as bovine spongiform encephalopathy–variant Creutzfeld–Jakob disease (BSE-vCJD), genetically engineered plants and dioxin contamination is still being experienced (Gale *et al.* 1998; Ferguson *et al.* 1999). One can only hope that the BSE-vCJD outbreak will catalyse improved, educated food safety systems that will prevent similar or even greater tragedies from occurring (Thomas & Newby 1999; Brown *et al.* 2001). An aspect that BSE-vCJD and publisied outbreaks of *E. coli* O157:H7 have emphasised is the need to control food safety 'from the farm to the fork', which is the approach adopted in this book.

The cost benefits in preventing food poisoning through the assured food safety system Hazard Analysis Critical Control Point (HACCP, Section 2.3) have been estimated (Crutchfield *et al.* 1999). Due to the range of economic models used, the estimates varied considerably from US$1.9 to US$171.8 billion. Regardless of the exact figure, it can be predicted that implementation of HACCP will reduce medical costs and productivity losses by an amount greater than the costs of implementing HACCP itself.

1.7   International control of microbiological hazards in foods

The increase in international trade in food has increased the risk from cross-border transmission of infectious agents, and underscores the need for international risk assessment to estimate the risk that microbial pathogens pose to human health. The globalisation and liberalisation of world food trade, while offering many benefits and opportunities, also presents new risks. Because of the global nature of food production, manufacturing and marketing, infectious agents can be disseminated from the original point of processing and packaging to locations thousands of miles away. Food-borne disease is of considerable importance as a global public health issue, and was recognised by the WHO as a priority area at the 53rd session of the World Health Assembly held in Geneva in May 2000. The Director General of the WHO, Dr Gro Harlem Brundtland, stated that there are three major challenges which need to be addressed to protect the health of the consumer (Brundtland 2001):

(1)   Re-establish consumer confidence from the farm to the table by reassessing and improving existing food safety systems
(2)   Ensure that reasonable food safety standards apply throughout the world and assist all countries to reach those standards
(3)   Develop global standards for pre-market approval systems of genetically modified food to ensure that these new products not only are safe, but also beneficial for consumers and more efficient than existing products.

The WHO and the Food and Agriculture Organisation (FAO) of the United Nations have been in the forefront of the development of risk-based approaches for the management of public health hazards in food. Because risk analysis is now well established for chemical hazards, the WHO and FAO are using their expertise for risk analysis of microbiological hazards (Chapter 3).

Conventional food safety systems, typified by the pasteurisation and sterilisation procedures of the dairy industry, have been improved by the adoption of HACCP (Section 2.3). Yet despite the increasing implementation of HACCP, the increased number of reported 'food poisoning' cases in many countries has underlined the need for further food safety systems, the emphasis being on assessing the direct microbiological risk to humans. This new approach, microbiological risk assessment, requires an improved epidemiological knowledge of food-borne diseases throughout the food chain. Because the food is in a global supply chain, this knowledge needs to be accumulated with a global perspective, thus involving governments as well as companies, which has resulted in the advent of the European Public Health and Food Safety Authority, and outbreak networks such as Enter-Net and Salm-Surv (see Internet Directory).

The 'farm to fork' approach to food safety has highlighted that it is easier to keep food products free from microbial (etc.) contamination in the supply chain if one can ensure contamination-free animals and poultry on the farm. An example is the approach in Sweden where *Salmonella*-free poultry production has almost been achieved. The programme costs about US$8 million per year to implement, but this amount is small compared to the cost of medical treatment which is estimated at US$28 million (Altekruse *et al.* 1993).

Antibiotics have been added to animal feeds to inhibit the growth of microbial pathogens. Unfortunately this has most probably caused the selection of antibiotic-resistant strains which can also be resistant to medically important antibiotics, with life-threatening consequences. Hence the use of antibiotics needs to be more carefully controlled (Section 4.3.4).

To ensure global food safety, developing and developed countries need to participate together in the establishment of food safety systems. WHO and its member states have recognized food safety as a world-wide challenge, and the World Health Assembly passed a food safety resolution in May 2000 recognising food safety as an essential public health issue. The resolution focuses on the need to develop sustainable, integrated food safety systems for the reduction of health risk along the entire food chain. Joint WHO/FAO activities are covered in greater detail in Section 3.2.4.

The WHO have given their aims for food safety in the twenty-first century as being:

- Strengthening national food safety policies and infrastructures
- Advising on food legislation and enforcement

- Evaluating and promoting safe food technologies
- Educating food handlers, health professionals and consumers in food safety
- Encouraging food safety in urban settings
- Promoting food safety in tourism
- Establishing epidemiological surveillance of food-borne diseases
- Monitoring chemical contamination of food
- Developing international food safety standards
- Assessing food-borne hazards and risks.

It is the last item on the WHO list which is the focus of this book.

The WHO aims to decrease the burden of microbiological food-borne diseases by taking into consideration the complete food chain and applying a holistic approach. This strategy comprises the following elements:

- Development of a strategy for microbiological monitoring in food (development of national capacities, investigation of potentials for database registration)
- Microbiological risk assessment methodology (development of international agreement on methodology, of guidance documents)
- Microbiological risk assessment advice (to the CAC and member countries, via joint expert consultations)
- Development and strengthening of risk communication methodology (guidelines for studying risk perception, methods for efficient communication, feasibility of a rapid alert system)
- New preventive strategies (potential of new food production, inspection or investigation methods to contribute food safety, expert consultation on emerging pathogens with priority given to food-borne viruses)
- Strengthening and co-ordination of global efforts on the surveillance of food-borne diseases and outbreak response (guidelines and standards for surveillance, laboratory and other capacities, networking).

Currently, food safety measures are not consistent around the world, and such differences can lead to trade disagreements among countries. This is particularly true if microbiological requirements are not justified scientifically. The standards, guidelines and recommendations adopted by the CAC and international trade agreements, such as those administered by the World Trade Organisation (WTO), are playing an increasingly important role in protecting the health of consumers and ensuring fair practices in trade. In 1962, the Joint FAO/WHO Food Standards Programme was created with the CAC as its executive organ. The *Codex*

Alimentarius, or the food code, is a collection of international food standards that have been adopted by the CAC. Codex standards cover all the main foods, whether processed, semi-processed or raw. The principal objectives of the CAC are to protect the health of consumers and ensure fair practices in the food trade (CCFH 1998).

In the case of microbiological hazards, the *Codex* elaborates standards, guidelines and recommendations that describe processes and procedures for the safe preparation of food. The application of these standards, guidelines and recommendations is intended to prevent or eliminate hazards in foods or reduce them to acceptable levels.

1.7.1   Sanitary and Phytosanitary (SPS) Agreement

The Final Act of the Uruguay Round of multilateral trade negotiations established the WTO to succeed the General Agreement on Tariffs and Trade (GATT). The Final Act led to the Agreement on the Application of Sanitary and Phytosanitary Measures (SPS Agreement) and the Agreement to Technical Barriers to Trade (TBT Agreement). The ratification of the WTO Agreement is a major factor in developing new hygiene measures for the international trade in food.

The GATT decision on SPS reaffirmed that no member state should be prevented from adopting or enforcing measures necessary to protect human, animal or plant life or health (GATT 1994). The SPS (Article 5) provisions of the WTO Agreement encourage 'harmonisation' of standards for food safety:

> Members shall ensure that their sanitary or phytosanitary measures are based on an assessment, as appropriate to the circumstances, of the risks to human, animal or plant health, taking into account risk assessment techniques as described by relevant international organisations.

These are intended to facilitate the free movement of foods across borders, by ensuring that means established by countries to protect human health are scientifically justified and are not used as non-tariff barriers to trade in foodstuffs. The Agreement states that SPS measures based on appropriate standards, codes and guidelines developed by the CAC are deemed to be necessary to protect human health and consistent with the relevant GATT provisions. The WTO SPS Agreement also recognizes the Office Internationale des Epizooties (OIE) and the International Plant Protection Convention (IPPC) as international standard-setting bodies for animal and plant health, respectively. The CAC, OIE and IPPC should co-ordinate their standard setting activities, as appropriate, to ensure that international food safety standards adequately consider and incorporate factors relevant to the impact of animal and plant health issues on food safety, e.g. BSE as a cause of illness (vCJD) in humans (Collinge *et al.* 1996).

The SPS Agreement came into force in 1995, in support of which the CAC now has a comprehensive action plan to incorporate risk analysis in its activities wherever appropriate. The SPS Agreement recognises the right of governments to protect their population's health from hazards that might be introduced through imported food by imposing sanitary measures, even though these might mean trade restrictions. However, such sanitary measures must be based on risk assessment in order to avoid unjustifiable, protective trade measures.

The management of food-borne hazards requires a transparent scientific risk assessment. Risk assessment needs to be undertaken by people that have scientific and public credibility and retain confidence in their conclusions. Both the risk assessment and risk management processes and decisions must be transparent and accompanied by effective communication activities. Many developing countries, however, are poorly equipped to manage effectively existing and emerging risks from food. As well as improved methods for the long-distance distribution of food, there have also been improvements in detection methods for pathogen food contaminants, both chemical and biological. However, the detected presence of a compound may not necessarily make it a public health hazard. Therefore risk assessment is necessary to assess the significance of the 'contaminant' and inform the public appropriately. Decisions must be made in accordance with internationally recognised standards which are scientifically and not politically based. Hence in the food industry, the *Codex* standards are internationally recognised and compliant with SPS provisions. Conventional Good Manufacturing Practice Based food hygiene requirements (i.e. end-product testing) are being improved through the implementation of HACCP. Subsequently risk assessment will put more emphasis on predictive microbiology for the generation of exposure data. In turn this will assist establishing Critical Limits for HACCP schemes (Buchanan 1995, 1997; Notermans *et al.* 1999; Serra *et al.* 1999) (Section 2.3).

The adoption of food safety systems such as HACCP has been encouraged by various international bodies such as the FAO, WHO and the CAC. Implementation of HACCP has changed the emphasis from reactive control of problems following end-product analysis to identifying hazards in the production process, and ensuring the steps which control the hazards are effective. Subsequently, HACCP has been adopted into law to different extents. However, HACCP does not assess the risk associated with food consumption as required by the SPS Agreement.

Hence the SPS Agreement provides an impetus for the development of microbiological risk assessment to support the elaboration of standards, guidelines and recommendations related to food safety. It provides a framework for the formulation and harmonisation of sanitary and phytosanitary measures. These measures must be based on science and implemented in an equivalent and transparent manner. They cannot be used as an unjustifiable barrier to trade by discriminating among foreign sources of supply or providing an unfair advantage to domestic producers. To facilitate safe food production for domestic and international markets, the SPS Agreement encourages governments to harmonise their national measures, or base them on international standards, guidelines and recommendations developed by international standard-setting bodies.

1.7.2 Origin of microbiological risk assessment

The purpose of the SPS and TBT Agreements is to prevent the use of national or regional technical requirements, or standards in general, as unjustified technical barriers to trade. The TBT Agreement covers all types of standards including quality requirements for foods (except requirements related to sanitary and phytosanitary measures), and it includes numerous measures designed to protect the consumer against deception and economic fraud. The TBT Agreement also places emphasis on international standards. WTO members are obliged to use international standards or parts of them except where the international standard would be ineffective or inappropriate in the national situation. The WTO Agreement also states that risk assessment should be used to provide the scientific basis for national food regulations on food safety and SPS measures, by taking into account risk assessment techniques developed by international organizations.

Because of the SPS Agreement and the WHO, the *Codex Alimentarius* standards, guidelines and other recommendations have become the baseline for safe food production and consumer protection. Hence the *Codex Alimentarius* has become the reference for international food safety requirements. The CAC has identified risk assessment of microbiological hazards in foods as a priority area of work. The Codex Committee on Food Hygiene (CCFH) is responsible for risk management of food in international trade because it has the overall responsibility for all provisions on food hygiene prepared by CAC commodity committees. CCFH identified a list of pathogen–commodity combinations on which expert risk assessment advice was required (see Table 5.1). Risk managers in CAC called for an expert advisory body regarding public health protection from food-borne hazard. Subsequently a joint FAO and WHO consultation on risk management and food safety concluded that the work of CCFH would benefit from advice from an expert body on food-borne microbiological hazards for purposes of risk management. The report suggested that such a committee of experts could provide scientific advice on microbiological risk assessment similar to that provided by the Joint FAO/WHO Expert Committee on Food Additives (JECFA) and the Joint FAO/WHO Meeting on Pesticide Residues (JMPR) on food additives, contaminants, veterinary drug residues and pesticide residues. The FAO and WHO are now co-operating to provide an expert based risk assessment of microbiological hazards in foods to their member countries and to the CAC (JEMRA). See Table 1.9 for a chronological listing of WHO and CAC activities.

Table 1.9 WHO/FAO Codex Alimentarius Commission (CAC) chronology of activities (see also Table 3.2).

1991–1993	Joint FAO/WHO conference on food standards, chemicals in food and food trade (1991) convened in collaboration with GATT. Recommends to CAC to incorporate risk assessment principles into decision making Codex 19th (1991) and 20th (1993) sessions agree on the incorporation of risk assessment principles in its procedures
1994	Codex executive committee urged FAO and WHO to convene a consultation on risk analysis
1995	Joint FAO/WHO expert consultation on the application of risk analysis to food standard issues. It recommended the basic terminology and principles of risk assessment and concluded that the analysis of risks associated with microbiological hazards presented unique challenges (FAO/WHO 1995) All relevant Codex committees to incorporate risk analysis concept into CAC procedures; FAO and WHO requested to convene further consultations on risk management and risk communications
1996	Codex Committee on Food Hygiene (CCFH) published *Principles and guidelines for the application of microbiological risk assessment*
1997	Joint FAO/WHO expert consultation on the application of risk management to food safety matters. The report identified a risk management framework and the elements of risk management for food safety. It described risk management as a continuing process of evaluation, option assessment, implementation, and monitoring and review (FAO/WHO 1997a) CAC decided to adopt key definitions of risk analysis terms related to food safety and to publish them in the Codex Alimentarius procedural manual (FAO/WHO 1997b)

(Contd)

Table 1.9   *(Contd)*

1998	Joint FAO/WHO expert consultation, *Application of risk communication to food standards and safety matters*, which identified elements and guiding principles of risk communication and strategies for effective risk communication (FAO/WHO 1998)
1999	Joint FAO/WHO expert consultation, *Risk assessment of microbiological hazards in foods*, aimed at developing an international strategy and supporting mechanisms for microbiological hazards in foods (FAO/WHO 1999) Codex Alimentarius Commission: *Principles and guidelines for the conduct of microbiological risk assessment* (CAC 1999)
2000	Preliminary document: *WHO/FAO guidelines on hazard characterization for pathogens in food and water* (WHO/FAO 2000a) Joint FAO/WHO expert consultation on *Interaction between assessors and managers of microbiological hazards in foods* (WHO/FAO 2000b) Joint FAO/WHO expert consultation, *Risk assessment of microbiological hazards in foods* (FAO/WHO 2000a) Joint FAO/WHO–CCFH document: *Draft principles and guidelines for the conduct of microbiological risk management* (CCFH 2000)

A foundation for microbiological risk assessment was established through a series of consultations held by FAO and WHO and through the documents developed by the CAC. Since 1995, WHO, in collaboration with FAO, has been developing a formalised framework of risk analysis of food-borne hazards. The process has included a number of joint expert consultations:

(1) The application of risk analysis to food standards and food safety issues which dealt principally with risk assessment, Geneva 1995
(2) Risk management and food safety, Rome 1997
(3) The application of risk communication to food standards and safety matters, Rome 1998
(4) Risk assessment of microbiological hazards in foods, Geneva 1999
(5) Risk management, Rome 2000
(6) The interaction between assessors and managers, Kiel 2000.

The principles and recommendations contained in the reports are intended to serve as guidelines for the CAC committees to review the standards and advisory texts in their respective areas of responsibility, and to provide a common framework to governments, industry and other parties wishing to develop risk analysis activities in the field of food safety. See the internet Directory for a listing of URLs from which the documents can be downloaded. The major publications from the CAC committees are:

(1) Principles and guidelines for the conduct of microbiological risk assessment, CAC/GL-30 (CAC 1999)
(2) Proposed draft principles and guidelines for the conduct of microbiological risk management (at step 3 of the procedure) CX/FH 00/6 (CCFH 2000).

The Joint FAO/WHO Expert Consultation on the Application of Risk Analysis to Food Standards Issues held in 1995 defined risk analysis as a process composed of three components (FAO/WHO 1995):

(1) Risk assessment
     A process of systematic and objective evaluation of all available information pertaining to food-borne hazards. Essentially this process identifies (and quantifies) the risk.
(2) Risk management
     The process of weighing policy alternatives in the light of the results of risk assessment and, if required, selecting and implementing appropriate control options, including regulatory measures. This process controls or prevents the risk.
(3) Risk communication
     The interactive exchange of information and opinions concerning risks and risk management among risk assessors, risk managers, consumers, and other interested parties. Hence this process informs others of the risk.

See Fig. 1.3.
     Risk assessment requires scientifically derived information and the application of established scientific procedures carried out in a transparent manner. However, sufficient scientific information is not always available and so an element of uncertainty must be associated with any decision.
     Risk management can be achieved through hygienic handling, processing and implementation of HACCP procedures, and the setting of criteria and/or standards.
     Risk communication provides the public with the results of expert scientific review of food hazards and their risk to the general public or specific groups, such as those who are immunodeficient, infants and the

Fig. 1.3   The general framework of microbiological risk assessment.

elderly. Risk communication provides industry and the consumer with information to reduce, prevent or avoid the food risk.

The details of these three components are still being debated and refined. In 1997 the European Union Scientific Co-operation Task (SCOOP, co-ordinated by France) was established on microbiological risk assessment for food-borne pathogens and toxins. It ran for 2 years and assessed the collation of microbiological risk assessment data within member countries of the European Union (SCOOP 1998). The study was the first of its kind to determine what different countries were doing with respect to microbiological risk assessment according to the definitions of the CAC (1998). The project involved 39 scientists from 13 countries. The study concluded that although microbiological risk assessment was rapidly developing across Europe, very few complete risk assessments had been published. Cassin *et al.* (1998a) introduced the term 'process risk model' which integrated microbiological risk analysis with scenario analysis and predictive microbiology. In addition, their use of the term 'dose–response assessment' is effectively equivalent to 'hazard characterisation' in the CAC and WHO 1995 terminology (WHO 1995; Potter 1996). The International Life Sciences Institute (ILSI 2000) have published a *Revised framework for microbial risk assessment* which is a revision of their

Food Safety Management Tools monograph (ILSI 1998a,b). Originally the ILSI risk assessment was more appropriate for water-borne microbial hazard (ILSI 1996). Although this is a very valuable document, the standard CAC approach will be used in this book with reference to ILSI where appropriate. A comparison of risk assessment synonyms is given in Fig. 1.4.

Therefore microbiological risk assessment has only relatively recently been applied to microbial food safety issues. According to Klapwijk *et al.* (2000), between 1994 and 1999 the majority of publications have come from the USA (24/66) and almost half the publications (35/66) have been reviews. There had been only seven full microbiological risk assessments which were focused on a specific pathogen-commodity combination:

(1)   Notermans *et al.* (1997): *B. cereus* risk in pasteurised milk.
(2)   Brown *et al.* (1998): *Salmonella* in chicken products
(3)   Cassin *et al.* (1998a): *E. coli* O157:H7 in ground (minced) beef
(4)   FSIS (1998): *S.* Enteritidis in shell eggs and egg products
(5)   FDA (1999): *L. monocytogenes* data survey
(6)   Soker *et al.* (1999): rotavirus in drinking water (ILSI format)
(7)   Teunis & Havelaar (1999): *Cryptosporidium* in drinking water (ILSI format).

A list of various pathogen–commodity studies is given in Table 1.10. These will be considered in greater detail in Chapters 3 and 4. Currently, the application of risk assessment to food microbiological safety is an evolving subject where areas lacking sufficient information are being identified for further research and analysis.

The management of microbiological hazards for foods in international trade can be divided into five steps (ICMSF 1997, 1998c):

(1)   Conduct a risk assessment
      The risk assessment and consequential risk management decisions provide a basis for determining the need to establish microbiological safety objectives; see Section 3.6.
(2)   Establish food safety objectives
      A microbiological food safety objective is a statement of the maximum level of a microbiological hazard considered acceptable for consumer protection. These should be developed by governmental bodies with a view to obtaining consensus with respect to a food in international trade.
(3)   Achievable food safety objectives
      The food safety objectives should be achievable throughout the food chain. This can be applied through the general principles of food

CAC	NAS-NRC RA	ILSI
Hazard identification The identification of biological, chemical, and physical agents capable of causing adverse health effects and which may be present in a particular food or group of foods	**Hazard identification** Determination of whether a specified chemical causes a particular health effect. Four classes of information used in this step are epidemiological data, animal-bioassay data, data on *in vitro* effects, and comparison of molecular structure	**Problem formulation** A systematic planning step that identifies the goals, breadth, and focus of the pathogen risk assessment, the regulatory and policy context of the assessment, and the major factors that need to be addressed for the assessment
Exposure assessment The qualitative and/or quantitative evaluation of the likely intake of biological, chemical and physical agents via food as well as exposures from other sources if relevant	**Exposure assessment** Determination of the extent of human exposure before or after application of regulatory controls	**Analysis phase** Technical examination of data concerning potential pathogen exposure and associated human health effects. Elements of this process are: **1. Characterisation of exposure** Evaluation of any interactions between the pathogen, the environment, and the human population. Steps include pathogen characterisation, pathogen occurrence and exposure analysis; the result is an exposure profile
Hazard characterisation The qualitative and/or quantitative evaluation of the nature of the adverse health effects associated with the hazard, including a dose–response assessment	**Dose–response assessment** Determination of the relationship between the magnitude of exposure and the probability of occurrence of the health effects in question. Methods include low-dose extrapolation and animal-to-animal extrapolation	**2. Characterisation of human health effects** Evaluation of the ability of a pathogen to cause adverse human health effects under a particular set of conditions. Steps include host characterisation, evaluation of human health effects and quantification of the dose–response relationship; the result is a host–pathogen profile
Risk characterisation The quantitative and/or qualitative estimation, including attendant uncertainties, of the probability of occurrence and severity of known or potential adverse health effects in a given population based on the above	**Risk characterisation** Description of the nature and often the magnitude of human risk, including attendant uncertainty	**Risk characterisation** Estimation of the likelihood of adverse human health effects occurring as a result of a defined exposure to a microbial contaminant or medium

Fig. 1.4 Synonymous terms for microbiological risk assessment (adapted from ILSI (2000)).

Table 1.10   Microbiological risk assessment studies (see also Table 3.2).

Micro-organism	References
Salmonella serovars Eggs and egg products	Todd (1996); Whiting & Buchanan (1997); FSIS (1998); Ebel *et al.* (2000)
Poultry industry	Brown *et al.* (1998); Oscar (1998a,b; see Internet Directory for Poultry FARM); Fazil *et al.* (2000a)
L. monocytogenes	Peeler & Bunning (1994); Farber *et al.* (1996); Van Schothorst (1996, 1997); Buchanan *et al.* (1997); Notermans *et al.* (1998a); Bemrah *et al.* (1998); FDA (1999); Todd *et al.* (1999); Lindqvist & Westöö (2000); FDA (2001)
E. coli O157:H7	Cassin *et al.* (1998a); Marks *et al.* (1998); Haas *et al.* (2000); Hoornstra & Notermans (2001)
Bacillus cereus	Zwietering *et al.* (1996); Todd (1996); Notermans *et al.* (1997); Notermans & Batt (1998); FSIS (1998); Carlin *et al.* (2000)

Campylobacter spp.
    Fluoroquinolone
       resistance
	Medema *et al.* (1996); Fazil *et al.* (2000b) FDA (2000a)

V. parahaemolyticus	FDA (2000b)
Sous-vide products	Barker *et al.* (1999)
BSE	Gale (1998)
Drinking water	FAO/WHO (2000b); Gale (2001)

hygiene and any product specific codes and HACCP systems. The HACCP requirements must be developed by the food industry; see Section 2.3.

(4)   Establish microbiological criteria, when appropriate
   This must be performed by an expert group of food microbiologists; see Section 2.8.

(5)   Establish acceptance procedures for the food at port of entry
   A list of approved suppliers as determined by inspection of facilities and operations, certification, microbiological testing and/or other testing such as pH and water activity measurements.

Therefore an understanding of HACCP, microbiological risk assessment (including food safety objectives) and microbiological criteria is required (Notermans *et al.* 1996). Microbiological risk assessment is only one integral component in a series of steps leading to the management of microbiological hazards for foods in international trade. To complete an effective microbiological risk assessment, it is imperative that key information about the food be available concerning the technologies and handling practices used from production to consumption.

2

FOOD SAFETY, CONTROL AND HACCP

2.1 Introduction

The microbiological safety of foods is principally assured by:

- Education and training of food handlers and consumers in the application of safe food production practices during production, processing (including labelling), handling, distribution, storage, sale, preparation and use
- Microbiological testing for the presence or absence of food-borne pathogens and toxins
- Implementation of Hazard Analysis Critical Control Point (HACCP)
- Control at the source
- Product design and process control.

Consideration of safety needs to be applied to the complete food chain. This is from food production on the farm (or equivalent) through to the consumer, commonly known as the 'from farm to fork' approach. To achieve this an integration of food safety tools is required (Forsythe & Hayes 1998, Fig. 2.1):

- Good manufacturing practice (GMP)
- Good hygienic practice (GHP)
- Hazard Analysis Critical Control Point (HACCP)
- Microbiological risk assessment (MRA)
- Quality management; ISO series
- Total quality management (TQM).

Because tools are implemented world-wide they will ease communication with food distributors and regulatory authorities, especially at port of entry (see Section 1.8).

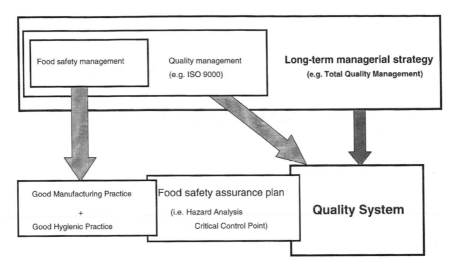

Fig. 2.1   Food safety management tools (adapted from ILSI 1998a).

2.2   HACCP adoption

Traditionally the safety of food being produced was assessed by end-product testing for the presence of food pathogens or their toxins. This *retrospective* approach, however, does not guarantee safe food. The world-wide accepted *proactive* approach is through the adoption of HACCP principles.

The HACCP system for managing food safety was derived from two major developments:

(1)   The HACCP system in the 1960s, as pioneered by the Pillsbury Company, the United States Army and NASA as a collaborative development for the production of safe foods for the USA space programme. NASA required 'zero defects' in food production to guarantee the astronauts' food (Bauman 1974).

(2)   Total quality management systems which emphasise a total system approach to manufacturing that could improve quality while lowering costs.

HACCP is a scientifically based protocol. It is systematic, identifies specific hazards and measures for their control to ensure the safety of food. It is interactive in that it involves the food plant personnel from those on the factory production line to the managers. HACCP is a tool to assess hazards and establish control systems. Its focus is on the prevention of problems occurring. HACCP schemes can accommodate change, such as advances in equipment design, processing procedures or technological developments (CAC 1997a,c).

HACCP can be applied throughout the food chain from primary production to final consumption, and its implementation should be guided by scientific evidence of risks to human health. Besides enhancing food safety, implementation of HACCP can provide other significant benefits. The application of HACCP systems can aid inspection by regulatory authorities and promote international trade by increasing confidence in food safety.

The successful application of HACCP requires the full commitment and involvement of management and the workforce. It also requires a multidisciplinary approach. The application of HACCP is compatible with the implementation of quality management systems, such as the ISO 9000 series, and is the system of choice in the management of food safety within such systems.

Guidance for the establishment of HACCP based systems is detailed in the publications *Hazard Analysis and Critical Control Point System and Guidelines for its Application* (CAC 1997a,c) and *HACCP in Microbiological Safety and Quality* (ICMSF 1988). Table 2.1 outlines the implementation of HACCP in a food company. The seven steps in bold are known as the 'seven principles of HACCP'.

Table 2.1 Establishing and implementing HACCP.

Decision by management to use the HACCP system
Training and formation of the HACCP team
Development of the HACCP plan document, including the following parts (CAC 1997a):
Assemble the HACCP team
Describe the food product and its distribution
Identify the intended use and consumers
Develop and verify the flow diagram for the production process
On-site confirmation of the flow diagram
(1) **Conduct a hazard analysis**
(2) **Determine the critical control points (CCPs)** (see Fig. 2.2)
(3) **Establish critical limits**
(4) **Establish monitoring procedures**
(5) **Establish corrective actions**
(6) **Establish verification procedures**
(7) **Establish documentation and record-keeping procedures**

2.3   Outline of HACCP

In order to produce a safe food product with negligible levels of food-borne pathogens and toxins, three control stages must be established:

(1) Prevent micro-organisms from contaminating food through hygienic production measures. This must include an examination of ingredients, premises, equipment, cleaning and disinfection protocols and personnel.
(2) Prevent micro-organisms from growing or forming toxins in food. This can be achieved through chilling, freezing, or other processes such as reduction of water activity or pH. These processes, however, do not destroy micro-organisms.
(3) Eliminate any food-borne micro-organisms, for example by using a time and temperature processing procedure or by the addition of suitable preservatives.

These principles were given by the CAC (1993) and the National Advisory Committee on Microbiological Criteria for Foods (NACMCF 1992, 1998a,b). Hence it is an internationally recognised procedure. Differences arise, however, with the interpretation and implementation of these seven principles. The approach here will adhere to the CAC, WHO and NACMCF (1998a,b) format. In the next section the *Codex Alimentarius* principles are given in italics. It should be noted that the *Codex Alimentarius* reverses Principles 6 and 7.

2.3.1   Food hazards

In HACCP a hazard is defined as

A biological, chemical, or physical property that may cause a food to be unsafe for consumption.

Biological hazards are living organisms, including microbiological organisms (bacteria, viruses, fungi) and parasites.

Chemical hazards are in two categories. The first group are naturally-occurring poisons, chemicals or deleterious substances. These are natural constituents of foods and are not the result of environmental, agricultural, industrial, or other contamination. Examples are aflatoxins and shellfish poisons. The second group are poisonous chemicals or deleterious substances which are intentionally or unintentionally added to foods at some point in the food chain. This group of chemicals can include pesticides and fungicides, as well as lubricants and cleaners.

A physical hazard is any physical material not normally found in food which causes illness or injury. Physical hazards include glass, wood, stones and metal which may cause illness and injury. Examples of hazards are given in Table 2.2.

Table 2.2   Hazards associated with food.

Biological	Chemical	Physical
Macrobiological	Veterinary residues:	Glass
Microbiological	antibiotics, growth	Metal
Viruses	stimulants	Stones
Pathogenic bacteria	Plasticisers and packaging	Wood
spore forming	migration: vinyl chloride,	Plastic
non-spore forming	bisphenol A	Parts of pests
bacterial toxins	Chemical residues:	Insulation
Shellfish toxins: domoic	pesticides (DDT), cleaning	material
acid, okadaic acid, NSP,	fluids	Bone
PSP[a]	Allergens	Fruit pits
Parasites and protozoa	Toxic metals: lead,	
Mycotoxins: ochratoxin,	cadmium, arsenic, tin,	
aflatoxins, fumonsins,	mercury	
patulin	Food chemicals:	
	preservatives, processing	
	aids	
	Radiochemicals: ^{131}I, ^{127}Cs	
	Dioxins, polychlorinated	
	biphenyls (PCBs)	
	Prohibited substances	
	Printing inks	

[a] NSP, neurotoxic shellfish poison; PSP, paralytic shellfish poison.
Adapted from Snyder (1995) and Forsythe (2000).

2.3.2   Pre-HACCP principles

Before the HACCP seven principles can be applied, there is the need to:

(1)   Assemble the HACCP team
      The food operation should ensure that the appropriate product-specific knowledge and expertise are available for the development of an effective HACCP plan. Optimally this may be accomplished by assembling a multidisciplinary team. Where such expertise is not available on site, expert advice should be obtained from other sources. The scope of the food chain is involved and the general classes of hazards to be addressed (e.g. does the plan cover all classes of hazards or only selected classes?).

(2) Describe the product

A full description of the product should be collated, including relevant safety information such as composition, physical/chemical structure (including a_w, pH, etc.), biocidal/static treatments (heat treatment, freezing, brining, smoking, etc.), packaging, durability, storage conditions and method of distribution.

(3) Identify intended use

The intended use should be based on the expected uses of the product by the end-user or consumer. In specific cases, vulnerable groups of the population, e.g. residents in institutions, may have to be considered.

(4) Construct flow diagram

The flow diagram should be constructed by the HACCP team. The flow diagram should cover all steps in the operation. When applying HACCP to a given operation, consideration should be given to steps preceding and following the specified operation.

(5) On-site confirmation of flow diagram

The HACCP team should compare the processing operation with the flow diagram during all stages and hours of operation, and amend the flow diagram where appropriate. The process operation should be checked on both day and night shifts as well as at weekends in case the personnel change the procedure according to the level of supervision.

2.3.3 Principle 1: Hazard analysis

Conduct a hazard analysis. Prepare a list of steps in the process where significant hazards occur and describe the preventative measures.

The HACCP team should list all the hazards that may be reasonably expected to occur at each step from primary production, processing, manufacture and distribution until the point of consumption. The evaluation of hazards should include the following:

• The likely occurrence of hazards and severity of their adverse health effects
• The qualitative and/or quantitative evaluation of the presence of hazards
• Survival or multiplication of micro-organisms of concern
• Production or persistence in foods of toxins, chemicals or physical agents
• Conditions leading to the above.

The hazard analysis should identify which hazards can be eliminated or reduced to acceptable levels as required for the production of safe food.

The common sources of food-borne pathogens are:

(1)   The raw ingredients
(2)   Personnel
(3)   The environment (air, water and equipment).

The severity of the illness caused by the organisms can be determined from standard texts, especially the ICMSF books, and is simplified in Table 1.3. The probable occurrence of the food-borne pathogen can also be determined from ICMSF and related literature (Table 1.2). A list of Web pages is given in the Internet Directory where further information on food-borne pathogens can be obtained; a standard reference source is the FDA Bad Bug Book (see Internet Directory). There are numerous Web sites giving information on current food poisoning outbreaks world-wide which may be accessed without any charges. National food poisoning statistics can be obtained from many regulatory authorities.

2.3.4   Principle 2: Critical Control Points

Identify the Critical Control Points (CCPs) in the process.

The HACCP team must identify the Critical Control Points (CCPs) in the production process which are essential for the elimination or acceptable reduction of the hazards that were identified in Principle 1. These CCPs are identified through the use of decision trees such as that given in Fig. 2.2 (NACMCF 1992). A series of questions are answered which lead to the decision as to whether the control point is a CCP. Other decision trees can be used if appropriate. The 'problem' with this principle is the term 'acceptable'. What is acceptable to the producer is probably different to what is acceptable to the consumer. This is where microbiological risk assessment should in the future assist in HACCP implementation, in particular in establishing CCPs which are scientifically justifiable and transparent.

A CCP must be a quantifiable procedure for measurable limits and monitoring to be achievable in Principles 3 and 4. If a hazard is identified for which there is no control measure in the flow diagram, then the product or process should be modified to include a control measure. Some groups used to classify CCPs into primary CCPs which eliminate hazards (CCP1s) and secondary CCPs which only reduce hazards (CCP2s). Although this approach is not currently recommended, it had the advantage of identifying which hazards are of crucial importance. For example, milk pasteurisation would be CCP1, whereas assessment of raw milk on receipt (for microbial load and antibiotic presence) would be CCP2 (Fig. 2.3). The

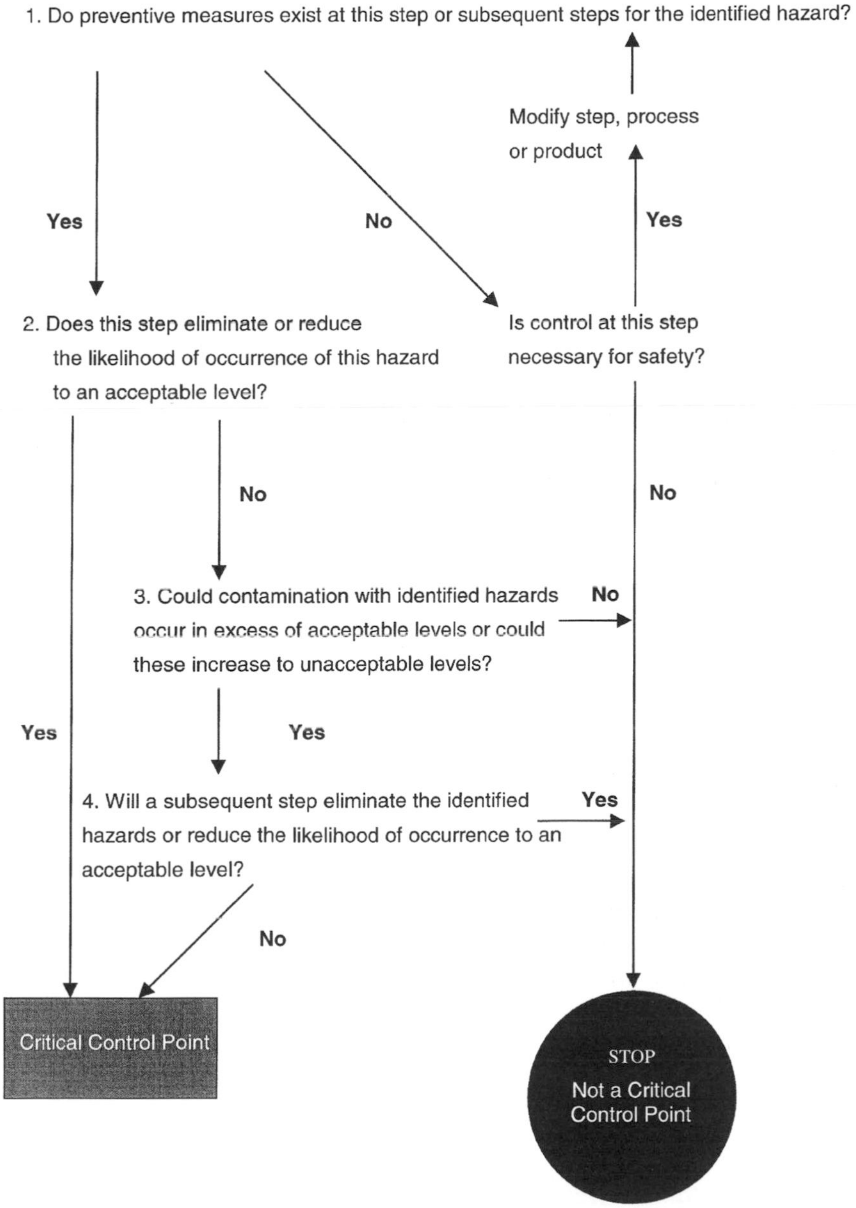

1. Do preventive measures exist at this step or subsequent steps for the identified hazard?

Modify step, process or product

Yes   **No**   **Yes**

2. Does this step eliminate or reduce the likelihood of occurrence of this hazard to an acceptable level?

Is control at this step necessary for safety?

No   **No**

3. Could contamination with identified hazards occur in excess of acceptable levels or could these increase to unacceptable levels?   **No**

Yes   **Yes**

4. Will a subsequent step eliminate the identified hazards or reduce the likelihood of occurrence to an acceptable level?   **Yes**

No

Critical Control Point

STOP
Not a Critical
Control Point

Fig. 2.2   Critical Control Point decision tree.

HAZARDS   **FLOW DIAGRAM**   **COMMENTS**

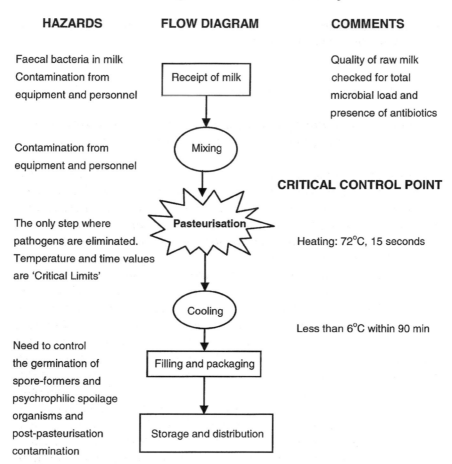

Faecal bacteria in milk

Contamination from

equipment and personnel

Receipt of milk

Quality of raw milk

checked for total

microbial load and

presence of antibiotics

Contamination from

equipment and personnel

Mixing

CRITICAL CONTROL POINT

The only step where

pathogens are eliminated.

Temperature and time values

are 'Critical Limits'

Pasteurisation

Heating: 72°C, 15 seconds

Cooling

Less than 6°C within 90 min

Need to control

the germination of

spore-formers and

psychrophilic spoilage

organisms and

post-pasteurisation

contamination

Filling and packaging

Storage and distribution

Fig. 2.3 HACCP flow diagram for the pasteurisation of milk.

cooking step is an obvious CCP for which Critical Limits of temperature and time can be set, monitored and corrected (Section 2.5.4). Non-temperature related control factors include water activity and pH (Section 2.5.1).

2.3.5 Principle 3: Critical Limits

Establish Critical Limits for preventative measures associated with each identified CCP.

Critical Limits must be specified and validated, if possible, for each CCP. The Critical Limit will describe the difference between safe and unsafe products at the CCP. A Critical Limit must be a quantifiable parameter such as temperature, time, pH, moisture or a_w, salt concentration or titratable acidity, available chlorine.

2.3.6 Principle 4: CCP monitoring

Establish CCP monitoring requirements. Establish procedures from the results of monitoring to adjust the process and maintain control.

Monitoring is the scheduled measurement or observation of a CCP relative to its Critical Limits. The monitoring procedures must be able to detect loss of control at the CCP. Monitoring should provide the information in time (ideally on-line) for correction of the control measure: by following the trend in measured values correction can take place before the value deviates from the Critical Limits.

The monitoring data must be evaluated by a designated person with knowledge and authority to carry out corrective actions when indicated. If monitoring is not continuous, then the amount or frequency of monitoring must be sufficient to guarantee the CCP is in control. Most monitoring procedures for CCPs will need to be done rapidly because they relate to on-line processes and there will not be time for lengthy analytical testing. Physical and chemical measurements are often preferred to microbiological testing, because they may be done rapidly and can often indicate the microbiological status of the product. Microbiological testing based on single samples or sampling plans will be of limited value in monitoring those processing steps which are CCPs (Section 2.8). This is mainly because the conventional microbiological methods are too time-consuming for effective feedback, and sampling plans have an inherent consumer–producer risk (Section 2.8.2). All records and documents associated with monitoring CCPs must be signed by the person(s) doing the monitoring and by a responsible reviewing official(s) of the company.

2.3.7 Principle 5: Corrective actions

Establish corrective actions to be taken when monitoring indicates a deviation from an established Critical Limit.

Specific corrective actions must be developed for each CCP to deal with deviations from the Critical Limits. The remedial action must ensure that the CCP is under control and that the affected product is recycled or destroyed as appropriate.

2.3.8   Principle 6: Verification

Establish procedures for verification that the HACCP system is working correctly.

Verification procedures must be established. These will ensure the HACCP plan is effective for the current processing procedure. NACMCF (1992) give four processes in the verification of HACCP:

(1)   Verification that critical limits at CCPs are satisfactory
(2)   Ensure that the HACCP plan is functioning effectively
(3)   Documented periodic revalidation, independent of audits or other verification procedures
(4)   Government's regulatory responsibility to ensure that the HACCP system has been correctly implemented.

The frequency of verification should be sufficient to confirm that the HACCP system is working effectively. Verification and auditing methods, procedures and tests, including random sampling and analysis, can be used to determine if the HACCP system is working correctly. Examples of verification activities include the following:

• Review of the HACCP system and its records
• Review of deviations and product dispositions
• Confirmation that CCPs are kept under control.

Verification should be conducted:

• Routinely, or on an unannounced basis, to assure CCPs are under control
• When there are emerging concerns about the safety of the product
• When foods have been implicated as a vehicle of food-borne disease
• To confirm that changes have been implemented correctly after a HACCP plan has been modified
• To assess whether a HACCP plan should be modified due to a change in the process, equipment, ingredients, etc.

2.3.9   Principle 7: Record-keeping

Establish effective record-keeping procedures that document the HACCP system.

HACCP procedures should be documented. Records must be kept to demonstrate safe product manufacture and that appropriate action has been taken for any deviations from the Critical Limits. Examples of documentation are:

- Hazard analysis
- CCP determination
- Critical Limits determination.

Examples of records are:

- CCP monitoring activities
- Deviations and associated corrective actions
- Modifications to the HACCP system.

2.4   Control at source

An initial approach to ensure the safety of food is to prevent contamination of the raw materials (CAC 1997b). For example, contamination of milk with *Brucella* spp. can be prevented by appropriate animal health measures (e.g. vaccination of animals). Levels of pesticide residues can be controlled by their proper application and good agricultural practice. Biotechnology can be used to produce plants which are more resistant to disease and thus require reduced use of pesticides. This may be particularly important in the case of production of plants which contain naturally occurring toxins. It should be noted that the perceived hazards of genetically modified foods are outside the scope of this book. For many chemical contaminants, such as PCBs and dioxins, preventing contamination through environmental measures may be the only practical means of keeping their concentrations at safe levels. The use of clean water for irrigation is important to reduce the transmission of hepatitis A virus. Preventing microbial contamination, however, is not always possible. The soil contains millions of bacteria, viruses and protozoa, including pathogenic varieties. Hence some potentially pathogenic micro-organisms are part of the natural flora of crops, animals and fish.

2.5   Product design and process control

Despite all efforts at good agricultural practice, certain raw materials may still become contaminated with pathogenic organisms. The hot and humid climates of many developing countries favour the growth of moulds and the production of mycotoxins. Improper handling during later stages in the food chain (i.e. transport, storage, distribution and preparation) may also increase the level of contaminants. Application of food technologies is thus essential to prevent food-borne diseases. In addition to controlling pathogens, most food technologies are also effective against spoilage micro-organisms and they are therefore often applied with the double objective of ensuring food safety and extending the shelf-life of food products.

From a public health viewpoint, food technologies may be divided into three groups, depending on their potential for preventing, reducing (killing) or controlling contaminants:

(1)   Technologies which render food safe (i.e. reduce contaminants present in food)
(2)   Technologies used to keep contaminants under control (i.e. prevent the growth of organisms or the production of toxins)
(3)   Technologies used to prevent contamination during or after processing.

In practice, combinations of two or more such technologies are frequently applied to achieve the desired objective. For example, raw milk is pasteurised to render it safe (see Fig. 2.3) and then it is aseptically packaged to prevent re-contamination.

Traditional ways to control microbial spoilage and safety hazards in foods include:

• Freezing
• Blanching
• Pasteurisation
• Sterilisation
• Canning
• Curing
• Syruping
• Inclusion of preservatives.

These food technologies broadly comprise those which entail a physical treatment (e.g. heating or freezing) and those which use chemical additives (e.g. curing). In the former, the foods are safe as long as surviving pathogens are kept under control and no contamination occurs after processing. In the latter, the control of contaminants persists as long as the chemical agents continue their activity in the food. To design adequate treatment processes an understanding of the factors affecting microbial growth is necessary.

2.5.1 Intrinsic and extrinsic factors affecting microbial growth

Food is a chemically complex matrix, and predicting whether or how fast micro-organisms will grow in any given food is difficult. Most foods contain sufficient nutrients to support microbial growth. Factors affecting microbial growth in food are divided into two groups of parameters, intrinsic and extrinsic (see Tables 2.3 and 2.4). Food additives, such as preservatives, are required to ensure that processed food remains safe and unspoiled during its shelf-life. A range of preservatives are used in food manufacture, including that of traditional foods. Many preservatives are effective under low pH conditions; benzoic acid (< pH 4.0), propionic acid (< pH 5.0), sorbic acid (< pH 6.5), sulphites (< pH 4.5). The parabens (benzoic acid esters) are more effective at neutral pH conditions. The various intrinsic and extrinsic parameters are described in greater detail in many other publications (see Forsythe 2000). For concise reference values on pH limits of microbial growth, the reader is directed to ICMSF (1996a) and related publications.

Table 2.3 Intrisic and extrinsic parameters affecting microbial growth.

Intrinsic parameters	Extrinsic parameters
Water activity, humectant identity	Temperature
Oxygen availability	Relative humidity
pH, acidity, acidulant identity	Atmosphere composition
Buffering capacity	Packaging
Available nutrients	
Natural antimicrobial substances	
Presence and identity of natural microbial flora	
Colloidal form	

2.5.2 Irradiation

The technology for food irradiation has been accepted in about 37 countries for 40 different foods. Twenty-four countries are using food irradiation at a commercial level. The standard covering irradiated food has been adopted by the CAC. It was based on the findings of a Joint Expert Committee on Food Irradiation (JECFI) which concluded that the irradiation of any food commodity up to an overall average dose of 10 kGy presented no toxicological hazard and required no further testing. The accepted dose (10 kGy) is the equivalent of pasteurisation and cannot sterilise food, nor inactivate food-borne viruses (Norwalk-like and hepatitis A). At low doses (up to 0.5 kGy), irradiation may be used to destroy

Table 2.4   Limits of microbial growth.

Organism	Mimimal water activity (a_w)	pH range	Temperature range (°C)[a]	D value
B. cereus	0.930	4.3–9.3	4–55	D_{100} 5.0
C. jejuni	0.990	4.9–9.5	30–45	D_{50} 0.88–1.63
Cl. botulinum type A and proteolytic B & F	0.935	4.6–9.0	10–48	$D_{121.1}$ 0.21
Cl. botulinum type E and non-proteolytic B & F	0.965	5.0–9.0	3.3–45	D_{82} 0.49–0.74
Cl. perfringens	0.945	5.0–9.0	10–52	D_{100} 0.3–20.0
E. coli O157:H7	0.935	4.0–9.0	7–49.4	D_{58} 1.6
L. monocytogenes	0.920	4.4–9.4	−0.4–45	D_{62} 2.9–4.2
S. Enteritidis	0.940	3.7–9.5	5–46	$D_{62.8}$ 0.06
St. aureus	0.830	4.0–10	7–50	$D_{65.5}$ 0.2–2.0
V. parahaemolyticus	0.936	4.8–11	5–44	D_{50} 0.9–1.1
Y. enterocolitica	0.945	4.2–10	−1.3–45	D_{60} 0.4–0.51

[a]To convert to degrees Fahrenheit use the equation $°F = (9/5)°C + 32$. As guidance: $0°C = 32°F$; $4.4°C = 40°F$; $60°C = 140°F$.
Various sources used, principally ICMSF (1996a), Corlett (1998) and Mortimore & Wallace (1994). Where data has differed between sources the wider growth range is quoted.

larval parasites in meat (e.g. *Trichinella spiralis* and *Taenia saginata*), or to inactivate metacercariae of *Clonorchis* and *Opisthorchis* in fish. The higher permitted dose (3–10 kGy) kills non-spore forming bacteria such as *Salmonella, Campylobacter* and *Vibrio* which may be present in meat, poultry and seafood. Irradiation can also be used to reduce the concentration of micro-organisms in spices and dried vegetables, and thus prevent contamination of foods to which these are added. Irradiation is a very useful pathogen control method for foodstuffs which are eaten raw or undercooked. Pork meat has been estimated to cause half to three-quarters of the infections with *Toxoplasma gondii* in the USA. Infections with protozoa and helminths are extremely common in certain tropical countries, and irradiation of meat and meat products can eliminate these risks. Similarly, irradiation can reduce the risk of poultry-borne salmonellosis or campylobacteriosis and the risk of *V. parahaemolyticus* gastroenteritis, salmonellosis and shigellosis transmitted through shrimps and seafood.

Irradiation technology can also be used as an alternative to chemical additives (which could be hazardous to human health) to reduce the level of spoilage micro-organisms, to delay sprouting or maturation and to disinfest grains. Irradiation is an alternative to fumigants such as ethylene bromide and ethylene oxide which have been banned or restricted for health or environmental reasons.

2.5.3 High pressure technology

The use of pressure technology is a novel means of food preservation (Knorr 1993; Hendrickx *et al.* 1998). The key objective is to reduce the number of food-borne pathogens by about 8 log cycles. Elevated pressure can exert detrimental effects on microbial physiology and viability. Growth of micro-organisms is generally inhibited at pressures in the range 20–130 MPa, while higher pressures of between 130 and 800 MPa may result in cell death; the maximum pressure allowing growth or survival depends on the species and medium composition. The sites of cell damage are possibly the cytoplasmic membrane and ribosomes.

The use of pressure technology has been proposed to inactivate bacterial spores. Spore germination is induced at low pressure (100–250 MPa) and is followed by inactivation at high pressure (500–600 MPa). A combination of pressure cycling and other preservation methods may be required for the control of *Clostridium* spp. because the spores of these organisms have greater pressure tolerance than those of *Bacillus* spp. Moderate hydrostatic pressure (270 MPa) at 25°C only reduces the viable count by about 1.3 log cycles. However 5 log cycle kills can be achieved by combined treatment conditions of 35°C and the inclusion of bactericidal compounds such as pediocin and nisin (Kalchayanand *et al.* 1998).

As the technology has little effect on chemical constituents, it keeps the natural flavour of food intact. In the future, this technology may provide a new method for rendering certain foods safe. Sometimes food technologies used for decontamination can be very simple. For example, the grating and mincing of cassava tubers during *gari* preparation is crucial for their detoxification.

2.5.4 Temperature control of microbiological hazards

For thousands of years, heat treatment has been the most effective and best method to control pathogenic and spoilage micro-organisms in food. Many food manufacturing processes include boiling, cooking or baking as heating steps. The resultant physicochemical changes increase the digestibility of certain foods as well as improving the food's texture, taste, smell and appearance. The need for adequate cooking for food safety, however, is often overlooked. It is therefore not surprising that inadequate temperature control is one of the major identified reasons for food poisoning outbreaks.

Food-borne pathogens grow within a specific temperature range which can be referred to as the 'danger' zone (Table 2.4). Outside this zone, growth is slowed down or stopped. For most pathogenic bacteria the danger zone is between 8 and 60°C. By keeping food at temperatures above 60°C (hot holding) or below 8°C (cold holding/chilling), it is possible to prevent or slow down the growth of most pathogenic bacteria or the production of toxins. Freezing is mainly used to prevent the growth of micro-organisms (both spoilage and pathogenic micro-organisms) as well as to maintain the quality of food. It also kills certain parasites such as *Trichinella* and *Taenia* in meat or *Anisakis* and *Clonorchis* in fish. In countries where it is a tradition to eat raw or undercooked fish or meats, for example, in China, Japan and Scandinavia, freezing can be an important technology for preventing food-borne helminthiasis.

Decimal reduction times (D values) and z values

To design an effective combined time and temperature treatment regime, it is imperative to understand the effects of heat on micro-organisms. The thermal destruction of micro-organisms (death kinetics of vegetative cells and spores) can be expressed logarithmically. In other words, for any specific organism in a specific substrate and at a specific temperature, there is a certain time required to destroy 90% (= 1 log reduction) of the organisms. This is the decimal reduction time (D value). A selection of D values is given in Table 2.4. Although most frequently applied to thermal death rates, D values can also be used to express the rate of death due to other lethal effects such as acid and irradiation.

The rate of thermal death depends upon the organism, including its ability to form spores, and the environment. Free (or planktonic) vegetative cells are more sensitive to detergents than fixed cells (biofilms or slime). The heat sensitivity of an organism at any given temperature varies according to the suspending medium. For example, the presence of acids and nitrite will increase the death rate whereas the presence of fat may decrease it. The D value is also dependent upon the inoculum preparation and the enumeration conditions. This has been demonstrated for *E. coli* O157:H7 and is summarised in Fig. 2.4 (Stringer *et al.* 2000). Hence D values quoted in books and journals cannot be taken as fixed values and directly applied to processes.

Plotting the D values for an organism in a substrate against the heating temperature should give a straight-line relationship (Fig. 2.5). The straight-line relationship does not always occur due to cell clumping. The slope of the line (in degrees Fahrenheit or Celsius) can be used to determine the change in temperature resulting in a ten-fold increase (or decrease) in the D value. This coefficient is called the z value. Hence if the organism has a z value of 5°C and a D value at 55°C of 10 minutes, then raising the treat-

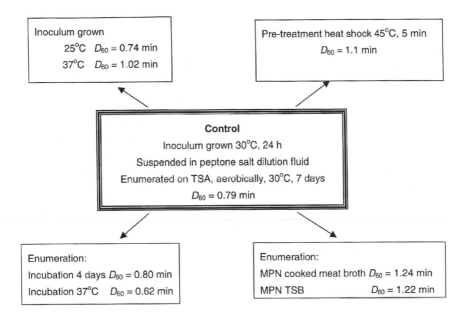

Fig. 2.4 Changes in *E. coli* O157:H7 D_{60} value (adapted from Stringer *et al.* (2000)).

ment temperature to 60°C means that the *D* value will be 1 minute, i.e. the organism will die ten times faster.

The cooking step is an obvious CCP in HACCP implementation for which Critical Limits of temperature and time can be set, monitored and corrected (Table 2.5). The time and temperature of the cooking process should be designed to give at least a 6 log kill of vegetative cells, i.e. 10^7 cells g^{-1} reduced to 10 cells g^{-1}. The treatment will not kill spores, and

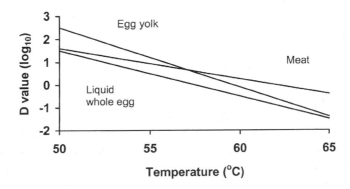

Fig. 2.5 D values for *S.* Enteritidis in eggs and meat (FSIS 1998; Fazil *et al.* 2000a).

Table 2.5   Effect of cooking temperature on the survival of *E. coli* O157:H7.

Temperature (°C)	Time for viable count to be reduced by 6 log cycles[a] (min)
58.2	28.2
62.8	2.8
67.5	0.28

[a] For example, from 10^7 cfu g^{-1} to 10 cfu g^{-1}. Data taken from $D_{62.8} = 0.4$, $z = 4.65$.

hence the time required to cool the food to a safe temperature needs to be monitored to prevent spore outgrowth.

The cooling period after heat treatment must be short enough to prevent spore outgrowth and germination from mesophilic *Bacillus* and *Clostridium* spp. Notably, the cooking process can create an anaerobic environment in the food which is ideal for *Clostridium* spp. The temperature range for growth of *Cl. perfringens* is 10–52°C (Table 2.4); hence cooling regimes should be designed to minimise the time the food is between these temperatures. A lower limit of 20°C is normally adopted because the organism only multiplies slowly below this temperature.

The control of the holding temperature of foods after processing and before consumption is crucial for safe food production. General recommended holding temperatures are:

- Cold served foods < 8°C
- Hot served foods > 63°C.

At cold serve temperatures (<8°C) *Clostridium* spp. are not a problem since they do not grow below 10°C. *Staphylococcus aureus* produces toxins above 10°C, but is able to multiply at temperatures down to 6.1°C. Subsequently temperature abuse can lead to *St. aureus* growth and toxin production. Since *St. aureus* toxin is not inactivated by reheating to 72°C, its production must be prevented because it cannot be inactivated by any further processing or cooking. Chilled foods should be stored below 3.3°C to ensure safety from spore germination and subsequent toxin production by *Cl. botulinum* E. In contrast, *Cl. botulinum* types A and B only multiply and produce toxin above 10°C. *Listeria monocytogenes* and *Y. enterocolitica* have a minimum growth temperature of −0.4°C and −1.3°C, respectively. Therefore the cold holding time must be limited for foods which are not going to be reheated. It should be noted that different national regulatory authorities will have different hot and cold holding temperatures. This illustrates the need for risks to be assessed on a worldwide basis.

2.6   Microbial response to stress

Micro-organisms are able, within limits, to adapt to stress conditions such as acidity and cold. The mechanism of adaptation is by signal transduction systems which control the co-ordinated expression of genes involved in cellular defence mechanisms (Huisman & Kolter 1994; Rees *et al.* 1995; Kleerebezem *et al.* 1997). Microbial cells are able to adapt to many processes used to retard microbial growth starvation, i.e. cold shock, heat shock, (weak) acids, high osmolarity and high hydrostatic pressure. The easiest mechanism of survival to recognise is that found in *Bacillus* and *Clostridium* spp. These organisms form spores under stress conditions which can germinate later under favourable conditions. Other organisms, such as *E. coli*, undergo significant physiological changes to enable the cell to survive environmental stresses such as starvation, near-UV radiation, hydrogen peroxide, heat and high salt concentrations. As already shown, due to adaptation the D_{60} value of *E. coli* O157:H7 is dependent upon the inoculum preparation and the enumeration conditions (Fig. 2.4).

2.6.1   *Response to pH stress*

Acidification is a commonly used method of preserving foods such as dairy products. Some micro-organisms, however, are able to tolerate and adapt to weak acid stresses, and this possibly enables them to survive passage through the acidic human stomach. Additionally, acid resistance can cross-protect against heat treatment, e.g. *E. coli* O157:H7 (Fig. 2.4 and Buchanan & Edelson 1999). The important aspect of the acid tolerance response is the induction of cross-protection to a variety of stresses (heat, osmolarity, membrane active compounds) in exponentially grown (log phase) cells. Acid-adapted cells are those that have been exposed to a gradual decrease in environmental pH, whereas acid-shocked cells are those which have been exposed to an abrupt shift from high pH to low pH. It is important to differentiate between these two conditions because acid-adapted, but not acid-shocked, *E. coli* O 157:H7 cells have enhanced heat tolerance in acidified tryptone soya broth at 52°C and 54°C, and in apple cider and orange juice (both of which are of low pH) at 52°C. Acid-induced general stress resistance may reduce the efficiency of hurdle technologies which are dependent upon multiple stress factors. Acid resistance can be induced by factors other than acid exposure (Baik *et al.* 1996; Kwon & Ricke 1998). *Salmonella* Typhimurium acid tolerance can be induced by exposure to short-chain fatty acids, which are used as food preservatives and also occur in the intestinal tract. Subsequently the virulence of *S.* Typhimurium may be enhanced by increasing acid resistance upon exposure to short chain fatty acids such as propionate, and further enhanced by anaerobiosis and low pH.

2.6.2   Heat- and cold-shock response

As previously stated, heat treatment is a common food preservation method. However, micro-organisms can adapt to mild heat treatment in a variety of ways.

- Cell membrane composition changes by increasing the saturation and the length of the component fatty acids to maintain the optimal membrane fluidity and the activity of intrinsic proteins
- There is accumulation of osmolytes that may enhance protein stability and protect enzymes against heat activation
- *Bacillus* and *Clostridium* spp. produce spores
- Heat-shock proteins (HSPs) are produced.

When bacterial cells are exposed to higher temperatures, a set of HSPs is rapidly produced. HSPs involve both chaperones and proteases which act together to maintain 'quality control' of cellular proteins. Production of HSPs is induced by several stress situations, e.g. heat, acid, oxidative stress and the presence of macrophages, suggesting that HSPs contribute to bacterial survival during infection. In addition, HSPs may enhance the survival of pathogens in foods during exposure to high temperatures. The process of adaptation and initiation of defence against elevated temperature is an important target when considering food preservation and the use of hurdle technology.

Cold adaptation by micro-organisms is of particular importance due to the increased use of frozen and chilled foods, and the increased popularity of fresh or minimally processed food containing few or no preservatives. Mechanisms of cold adaptation are as follows:

- Membrane composition modifications, to maintain membrane fluidity for nutrient uptake
- Structural integrity of proteins and ribosomes
- Production of cold shock proteins (CSPs)
- Compatible solutes.

To maintain membrane fluidity and function at low temperatures, micro-organisms increase the proportion of shorter and/or unsaturated fatty acids in their lipids. Cold shock proteins are small (7 kDa) proteins which are synthesized when bacteria are subjected to a sudden decrease in temperature. CSPs are involved in protein synthesis and mRNA folding.

Different cold-shock treatments before freezing result in differences in survival of bacteria after freezing. This might result in high survival of bacteria in frozen food products. Furthermore, low temperature adapted bacteria may be relevant to food quality and safety.

2.6.3 Response to osmotic shock

Lowering water activity (a_w) is one of the common ways of preserving food. The microbial response to loss of turgor pressure is the cytoplasmic accumulation of 'compatible solutes' that do not interfere too seriously with cellular functions (Booth *et al.* 1994). These compounds are small organic molecules, which share a number of common properties:

- Soluble to high concentrations
- Can be accumulated to very high levels in the cytoplasm
- Neutral or zwitterionic molecules
- Specific transport mechanism present in the cytoplasmic membrane
- Do not alter enzyme activity
- May protect enzymes from denaturation by salts or protect them against freezing and drying.

The adaptation of *E. coli* O 157:H7, *S.* Typhimurium, *B. subtilis*, *St. aureus* and *L. monocytogenes* to osmotic stress is mainly through the accumulation of betaine (*N*, *N*, *N*-trimethylglycine) via specific transporters. Other compatible solutes include carnitine, trehalose, glycerol, sucrose, proline, mannitol, glucitol, ectoine and small peptides.

2.6.4 Response to high hydrostatic pressure

The use of high hydrostatic pressure is a novel means of food preservation (Section 2.5.3). Exposure of *E. coli* to high pressure results in the synthesis of HSPs, CSPs and proteins only associated with high-pressure exposure.

2.7 Predictive modelling

Predictive food microbiology is a field of study that combines elements of microbiology, mathematics and statistics to develop models that describe and predict the growth and decline of microbes under prescribed environmental conditions (Whiting 1995). Obviously, if food is subjected to temperature fluctuations during distribution and storage then the rate of microbial growth will be affected. There are 'kinetic' models which model the extent and rate of microbial growth and 'probability' models which predict the likelihood of a given event occurring, such as sporulation (Table 2.6; Ross & McMeekin 1994).

The origin of predictive microbiology is in the canning industry and the *D* values used to describe the rate of microbial death. However, the ability to solve complex mathematical equations has required the revolution in computing power, and this has greatly facilitated the development of predictive modelling. Models developed to predict microbial survival and growth may become an integral tool to evaluate, control, document and even defend the safety designed into a food product (Baker 1995). A form of predictive microbiology is the development of expert systems that can quantify the safety risks of food products without the necessity of extensive laboratory work (Schellekens *et al.* 1994; Wijtzes *et al.* 1998).

The main objective of predictive microbiology is to describe mathematically the growth of micro-organisms in food under prescribed growth conditions. The major factors affecting microbial growth are:

Table 2.6 Examples of predictive modeling of microbial growth and toxin production.

Organism	Comments	Reference
B. cereus		Zwietering *et al.* (1996)
Brochothrix thermosphacta		McClure *et al.* (1993)
Cl. botulinum types A & B	Growth and toxin production	Robinson *et al.* (1982), Roberts & Gibson (1986), Lund *et al.* (1990)
Cl. botulinum	Toxin production	Hauschild *et al.* (1982), Lindroth & Genigeorgis (1986), Baker & Genigeogis (1990) Meng & Genigeorgis (1993)
E. coli O157:H7		Sutherland *et al.* (1995)
St. aureus		Broughall *et al.* (1983), Sutherland *et al.* (1994)
S. Typhimurium		Broughall & Brown (1984), Thayer *et al.* (1987), Oscar (1998a,b, 1999a–c)
Salmonella spp. (mixed)		Gibson *et al.* (1988)
Yersinia enterocolitica		Sutherland & Bayliss (1994)

- pH
- Water activity
- Atmosphere
- Temperature
- Presence of certain organic acids, such as lactate.

Initially the data for predictive modelling is collected using a range of bacterial strains to represent the variation of the target organism present in the commercial situation. Ideally this will include strains associated with outbreaks, the fastest growing strains and the most frequently isolated strain. Although there is a considerable wealth of knowledge from the growth of microbes in bioreactors (fermenters), this is not directly applicable to the food industry. In food, the environmental factors are more varied and fluctuate, and one is often dealing with a mixed population. Detailed studies of predictive microbiology are given by McMeekin *et al.* (1993) and McDonald & Sun (1999).

Predictive growth models can be thought of as having three levels (Whiting 1995; McDonald & Sun 1999):

(1) Primary level models describe changes in microbial numbers (or equivalent) with time
(2) Secondary level models show how the parameters of the primary model vary with environmental conditions
(3) Tertiary level models combine the first two types of model with user-friendly application software or expert systems that calculate microbial behaviour under changing environmental conditions.

2.7.1   *Primary models and the Gompertz and Baranyi equations*

Primary models may quantify the increase in microbial biomass as colony forming units per millilitre, or as absorbance. Alternatively, quantification could be in terms of changes in media composition, such as metabolic end-products, conductivity or toxin production. Early primary models describing growth parameters started by plotting the growth curve and determining the rate of growth from the exponential phase. Further models were developed later which included the lag phase.

Gompertz model
The Gompertz function has become the most widely used primary model (Pruitt & Kamau 1993). It can be represented as

$$\log(N_t) = A + C \exp(-\exp(\exp(-B(t - M))))$$

where N_t is the population density (cfu ml^{-1}) at time t (hours), A is the logarithm of initial population density [log(cfu ml^{-1})], C is the logarithm of the difference in initial and maximum population densities [log(cfu ml^{-1})], M is the time of maximum growth rate (hours), and B is the logarithm of the relative maximum growth rate at M [log(cfu ml^{-1}) hour^{-1}]. This Gompertz function produces a sigmoidal curve that consists of four phases comparable to the phases of microbiological growth: lag, acceleration, deceleration and stationary.

Baranyi model

Baranyi and co-workers developed an alternative equation based on the basic growth model which incorporated the lag, exponential and stationary phases and the specific growth rate:

$$N_t = N_{max} - \ln[1 + (\exp(N_{max} - N_0) - 1\exp(-\mu_{max}A(t)]$$

where N_t is the logarithm of the population size at time t, N_o is the logarithm of the initial population size, N_{max} is the logarithm of the maximum population, μ_{max} is the maximum specific growth rate and $A(t)$ is the integral of the adjustment function at time t. The model fitted experimental data better than the Gompertz function with regard to predicting the lag time and exponential growth phase.

The Baranyi equation has been used as the basis (with modification) for the MicroFit software. MicroFit is a freeware program (see Internet Directory for the downloading address) developed by MAFF (UK) and four other partners. Microbiological data is analysed is according to the Baranyi growth model (Baranyi & Roberts 1994). The program enables microbiological growth data to be easily analysed to

- determine μ_{max}, doubling time, lag time, initial cell count and final cell count
- estimate confidence intervals on the above parameters
- simultaneously analyse two datasets and compare them graphically
- perform statistical testing on the difference between two datasets.

The model cannot analyse the decline phase because this is not described by the Baranyi growth model.

2.7.2   Secondary models

Commonly used secondary models describe the responses to changes in an environmental factor such as temperature, pH and a_w. There are three types of model: second-order response surface equation; the square root model (Belehardek); and Arrhenius relationships. Broughall *et al.* (1983) used the Arrhenius equation to describe the lag and generation time of *St.*

aureus and *S.* Typhimurium. The square root model has been applied to the growth of *E. coli*, *Bacillus* spp., *Y. enterocolitica* and *L. monocytogenes* (Gill & Phillips 1985; Adams *et al.* 1991; Wimptheimar *et al.* 1990; Heitzer *et al.* 1991). To have sufficient data for model fitting, a large number of data points must be collected. Growth can be measured by a number of methods such as turbidity and plate counts. Methods of automatic data collection are of obvious benefit because they are less laborious and the data is digitised. The Bioscreen (Labsystems) automatically records the turbidity of a large number of samples at any given time. Borch & Wallentin (1993) used conductance (impedance) microbiology to model the growth rate of *Y. enterocolitica* in pork and the data closely fitted the Gompertz function.

2.7.3 Tertiary models

Tertiary models use the primary and secondary model to generate models used to calculate the microbial response to changing conditions and to compare the effects of different conditions. The Pathogen Modeling Program and the FoodMicromodel program are two easily available tertiary models. The Pathogen Modeling Program was developed by the USDA Food Safety Group as a spreadsheet software-based system (see Internet Directory for the downloading address). It includes models for the effect of temperature, pH, water activity, nitrite concentration and atmospheric composition on the growth and lag responses of major food pathogens (Fig. 2.6). The Food Micromodel was developed in the UK by

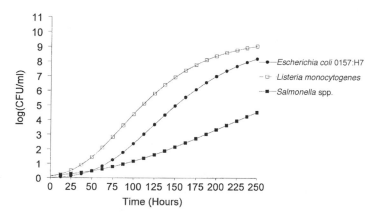

Fig. 2.6 Pathogen Modeling Program simulation for the growth of *E. coli* O157:H7, *L. monocytogenes* and *Salmonella* spp. at 10°C, pH 6.5 and 0.9% NaCl.

the Campden Food and Drink Research Association and includes environmental parameters similar to the Pathogen Modeling Program (Jones 1993). It has predictive equations for growth, survival and death of pathogens.

Predictive models are available for use in HACCP and microbiological risk analysis (Van Gerwen & Zwietering 1998). Models can be used to assess the risk of probability and determine the consequence of a microbiological hazard in food (see Chapter 4 for detailed examples). By using predictive models, ranges and combinations of process parameters can be established as Critical Limits for CCPs. Predictive microbiological models are tools to aid the decision-making processes of risk assessment and to describe the process parameters necessary to achieve an acceptable level of risk (Elliott 1996). Using a combination of risk assessment and predictive modelling it is possible to determine the effects of processing modifications on food safety. Figure 2.7 shows the predicted effect of temperature abuse on the probability of infection (Section 3.3.5) for *Sh. flexneri*. The Pathogen Modeling Program was used to predict the rate of *Sh. flexneri* growth under defined conditions (pH, salt, temperature) for increasing time-periods and the subsequent likelihood of infection (Buchanan & Whiting 1998).

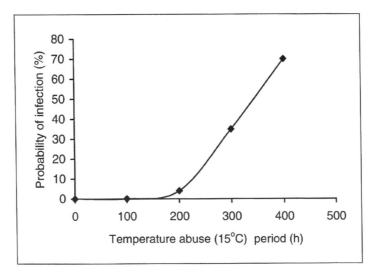

Fig. 2.7 Predicted effect of temperature abuse duration on *Sh. flexneri* probability of infection at pH 6.5, NaCl 0.5% calculated using the Pathogen Modeling Program (from Buchanan & Whiting (1996)).

2.8 Microbiological criteria

The ICMSF book *Microorganisms in Foods, Vol 2. Sampling for Microbiological Analysis: Principles and Specific Applications* was first published in 1974 and republished in 1986. It recognised the need for scientifically based sampling plans for foods in international trade. The sampling plans were originally designed for application at port of entry, i.e. when there is no prior knowledge on the history of the food (European Commission 1999). This pioneering work set forth the principles of sampling plans for the microbiological evaluation of foods. It is also known as attributes and variables sampling, depending on the extent of microbiological knowledge of the food. Subsequent books were published by the ICMSF to assist in the interpretation of microbiological data, such as *Microbiological Ecology of Food Commodoties* (ICMSF 1998a), *Characteristics of Microbial Pathogens* (ICMSF 1996a). The second edition of *Sampling for Microbiological Analysis* (ICMSF 1986) took note of the successful application of the acceptance sampling plans on a worldwide basis, not only at an international level but at national and local levels by both industry and regulatory agencies. Additionally, Harrigan & Park (1991) wrote an excellent book on the practical mathematics of sampling plans.

The purpose of establishing microbiological criteria is to protect the public's health by providing food which is safe, sound and wholesome, and to meet the requirements of fair trade practices (see Section 1.8.1). In time, the microbiological criteria may be set from food safety objectives (Section 3.5). The presence of criteria, however, does not protect the consumer's health because it is possible for a food-lot to be accepted which contains defective units. Microbiological criteria may be applied at any point along the food chain, and can be used to examine food at the port of entry and at the retail level. The microbiological analysis of food using sampling plans results in the acceptance or rejection of a food-lot. This includes the statistical probability that a food-lot may be falsely accepted or rejected, which is known as 'producer's risk' and 'consumer's risk'. Hence statistical analysis demonstrates that microbiological testing is not a stand-alone verification tool for hazard analysis systems (see HACCP Principle 6, Section 2.3.8).

2.8.1 Attributes sampling plan

Just as it is impractical to test a sample for every possible food pathogen, so it is also impractical to test 100% of an ingredient or end-product. Therefore there is a need to use sampling plans to appropriately test a batch of material and give a statistical basis for acceptance or rejection of a

- The probability of acceptance (P_a) on the *y*-axis, where P_a is the expected proportion of times a food-lot of this given quality is sampled for a decision
- Percentage defective sample units comprising a lot (*p*) on the *x*-axis; this is also known as a measure of food-lot quality.

Figure 2.8 gives the operating characteristic curve for the sampling plan $n = 5$, $c = 1$–3. This curve emphasises the high chance of accepting lots with up to 30% defectives. If a producer sets a limit of 10% defectives (i.e. $p = 10\%$) using a two-class plan of $n = 5$ and $c = 2$, then the probability of acceptance P_a is 99%. This means that on 99 of every 100 occasions when a 10% defective lot is sampled, one may expect to have two or fewer of the five tests showing the presence of the organism and thus calling for 'acceptance', while on one of every 100 occasions there will be three or more positives, calling for non-acceptance. Therefore a sampling plan of $n = 5$ and $c = 2$ will mean that a 10% defective lot will be accepted on most (99%) sampling occasions! Even increasing the number of samples to ten ($n = 10$, $c = 2$) means that 10% of defective batches will be accepted on 93% of occasions. Therefore no sampling plan can guarantee the absence of a pathogen, unless every gram of the food was to be analysed leaving nothing for consumption. Hence there is an essential need for the proactive approach of HACCP for assured food safety (Section 2.3).

The operating characteristic curve shows that it is possible that a 'bad' lot of food will on occasions be accepted, and conversely a 'good' lot will be rejected. This is known as the 'consumer's risk' and 'producer's risk', respectively. The 'consumer's risk' is considered to be the probability of

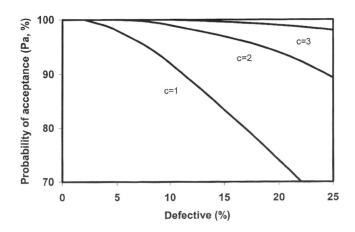

Fig. 2.8   Operating characteristic curve for $n = 5$ and $c = 1$–3.

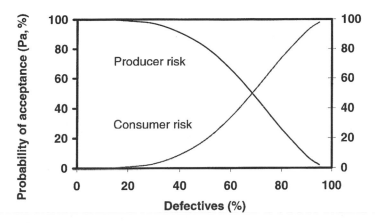

Fig. 2.9 Producer risk/consumer risk curve.

accepting a food-lot the actual microbial content of which is substandard as specified in the plan, even though the microbiological analysis of the sample units conforms to acceptance (P_a). The 'producer's risk' is expressed by '$1 - P_a$' (Fig. 2.9).

3

RISK ANALYSIS

3.1 Introduction

Definition: Risk is a function of the probability of an adverse health effect and the severity of that effect, consequential to a hazard(s) in food.

Risk analysis is the 'third wave' of food safety (Schlundt, Risk assessment advice, WHO online training course, see Internet Directory), the three being

(1) Good hygienic practices in production and preparation to reduce the prevalence and concentration of the microbial hazard
(2) HACCP and HACCP-like approach which proactively identifies and controls the hazard
(3) Risk analysis which focuses on the consequences to humans of ingesting the microbial hazard, and the occurrence of the hazard in the whole food chain (from farm to fork).

As described in Chapter 1, changes in food processing techniques, food distribution and the emergence of new food-borne pathogens can change the epidemiology of food-borne diseases. The increase in international trade in food has increased the risk from cross-border transmission of infectious agents and underscores the need to use international risk assessment to estimate the risk that microbial pathogens pose to human health. The globalisation and liberalisation of world food trade, while offering many benefits and opportunities, also presents new risks. Because of the global nature of food production, manufacturing, and marketing, infectious agents can be disseminated from the original point of processing and packaging to locations thousands of miles away. Therefore new strategies are required for evaluating and managing food safety risks. Risk analysis generates models which will enable the changes in food processing, distribution and consumption to be assessed with regard to their influence on food poisoning potential. It is a management tool, initially for governmental bodies by which to define an appropriate level of protection and establish guidelines to ensure the supply of safe foods. Nevertheless, it is also a tool for food companies with which they can assess the effect of processing changes on microbial risks (risk mitigation strategies).

Food safety must be ensured by the proper design of the food product and the production process. This involves combined control through intrinsic and extrinsic parameters (Table 2.3) as well as conditions for handling, storage, preparation and use. The current assured method of controlling hazards is HACCP (Section 2.3), whereas microbiological risk assessment is the stepwise analysis of hazards that may be associated with a particular type of food product, permitting an estimation of the probability of occurrence of adverse effects on health from consuming the product in question (Notermans & Mead 1996). It is sometimes referred to as Quantitative Microbial Risk Assessment (Haas *et al.* 1999) and can be described as a methodology to organise and analyse scientific information to estimate the probability and severity of an adverse event (Cassin *et al.* 1998b).

Although many food-borne micro-organisms have been recognised as pathogenic (Table 1.1), not every ingestion of a pathogen results in an infection or subsequent illness. There is variation in the infectivity of the micro-organism as well as variation in the population susceptibility. Therefore the risk of a food-borne disease is the combination of the likelihood of exposure to the pathogen through ingestion and the likelihood that the exposure will result in infection/intoxication and subsequent illness, of which there are varying degrees including death. The risk can be quantified on a population basis to predict the likely number of infections, illnesses or deaths per 100 000 population per year, or per meal, etc. It can also help to identify those stages from 'farm to fork' that contribute to an increased risk of food-borne illness, and help focus on steps that most effectively reduce the risk of food-borne pathogens. These are termed 'risk mitigating strategies'. Where insufficient data are available, qualitative (descriptive) risk assessments can be carried out. A qualitative risk assessment may be first constructed before deciding if a more time-consuming, resource-demanding quantitative risk assessment is necessary.

Therefore risk assessment has two main objectives:

(1) Quantify the risk to a defined population group from consumption of a defined product. If there are sufficient data, determine the risk from levels and frequency of contamination at the time of consumption, the amount of consumption (meal size and frequency) and an appropriate dose–response relationship to translate the exposure into public health outcomes.

(2) Identify strategies and actions that can be used to decrease the level

of health risk. This usually requires the modelling of production, processing and handling of the food, and changes in the 'farm to fork' chain. Subsequently it may identify the steps in food production that are critical to food safety, and those at which control actions or interventions would produce the greatest reduction in risk of food-borne illness. Hence it is of potential use for CCP identification for HACCP implementation.

As previously described, risk analysis consists of three components:

(1) Risk assessment: identifies the risk and factors that influence it
(2) Risk management: shows how can the risk be controlled or prevented
(3) Risk communication: informs others of the risk.

Although these three activities must be kept separate, there is, nevertheless, an essential exchange of information between them (Fig. 1.3). Risk analysis should be carried out in accordance with the internationally accepted principles established by the FAO/WHO. These principles form the basis of a Codex Committee on Food Hygiene document on risk management, which is currently in draft form (CCFH 2000).

The risk analysis starts with the gathering and brief description of a profile of the current knowledge on the micro-organism in question. This includes its incidence in the environment, in certain domestic animals and foods, as well as the implications for humans. The risk profile can then form the basis of the next step in the risk analysis, which is a scientific risk assessment of the micro-organism using scientific literature. During this time new research into poorly understood areas may be initiated. Relevant data regarding the seriousness and incidence of the micro-organism will be subjected to further analysis by means of models that demonstrate the significance of food contamination to human illness and its severity. On the basis of this, an assessment will be presented of how serious is the threat to people's health actually posed by the micro-organism. At the same time, the risk assessment will provide information regarding the most effective measures taken to prevent the existence of the micro-organism in foods.

The risk analysis will provide the information for defining an acceptable level of the pathogenic micro-organism (limit value) or an acceptable prevalence in a product, which can subsequently be achieved by the implementation of plans for prevention, HACCP-plans, rules for hygiene, etc.

Safety concepts need to be built into the development of food products, for example through HACCP implementation. Subsequently these must be incorporated into Good Manufacturing Practices, Good Hygienic Practices and Total Quality Management (Fig. 2.1). Hence, in the future, microbiological risk assessment should provide better information for the development of HACCP schemes, especially setting CCPs and company food safety activities (Fig. 3.1). However, it can take several years for a formal risk analysis to be completed. To assist food companies, the WTO allows internationally accepted criteria to be used. These criteria must be based on previous risk analysis procedures with the CAC and national governments as the risk managers. When the food safety objectives (Section 3.6) have been defined, the food companies will need to convert them into their own product or process criteria (Fig. 3.1). Because of limited resources, food companies will take a simpler approach than governmental risk managers. They will focus on the prevalence and concentration of a recognised pathogen in their food ingredients and finished product (see Section 3.3.3 on exposure assessment) and can use predictive microbiology to determine the likely changes in prevalence and concentration during processing, preparation and storage. Subsequently, they can identify factors (i.e. initial microbial load) which could be controlled to reduce the associated risk (risk mitigation). See Fig. 2.7 for an example using *Sh. flexneri* (Section 2.7.3).

3.2   Overview of microbiological risk assessment

The National Academy of Sciences initially developed the procedures for risk analysis (National Academy of Sciences, USA 1983; NRC 1993, 1994, 1998) which have been adapted in the *Codex Alimentarius*. Since 1995, WHO and FAO have been developing a formalised framework of risk analysis of food-borne hazards (see Section 1.8). The National Advisory Committee on Microbiological Criteria for Foods has published generic principles of risk assessment for illnesses caused by food-borne hazards (NACMCF 1998a). Risk assessment has been defined for microbiological hazards in foods by the 32nd meeting of the CAC (1999). See Fig. 1.4 for a comparison of terms between organisations.

Compared with risk assessment of chemical hazards, the risk assessment of biological hazards is a 'new developing' science. Microbial risk assessment varies from chemical risk assessment (ILSI (2000); see also Table 3.1) in the following ways:

(1)   Micro-organisms may multiply or die in food, whereas post-harvest concentrations of chemicals in edible animal products do not change very much.

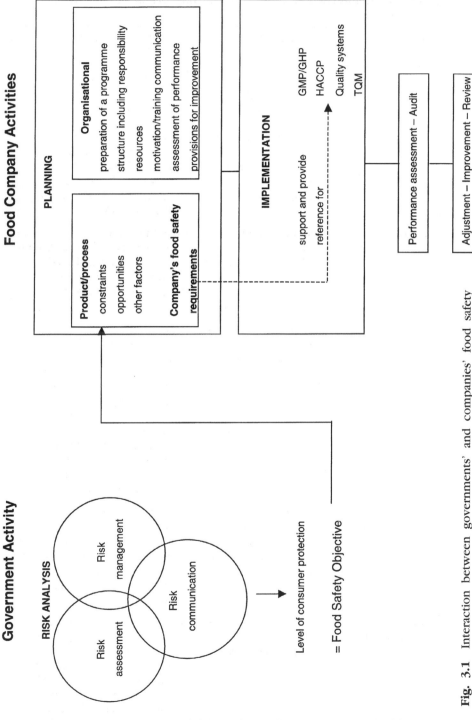

Fig. 3.1 Interaction between governments' and companies' food safety activities (adapted from ILSI (1998a)).

Table 3.1 Comparison of microbiological risk assessment with chemical risk assessment.

Component	Chemical risk assessment	Microbial risk assessment
Purpose	Determine: (a) If drug is genotoxic (b) Estimate safe levels in foods (c) Drug withholding times Provides for drug use that poses insignificant health risk	Characterise risk in terms of the probability of illness or death from naturally occurring microbial contaminants Assess impact of changes in food production and/or processing on risk Develop food standards Eventually establish critical limits for HACCP
Hazard identification	Chemical structure of drug Evidence of toxicity in animal bioassay and no observed effect level (NOEL) (dose) derived	Agent identified Evidence of causal role in food-borne disease gathered from outbreak investigation or epidemiological study
Exposure assessment	Assumptions made about consumption of foods derived from treated animals With ADI calculate maximum residue limits (MRL) in foods Withdrawal time set to ensure MRL not exceeded	Usually complex determinations of prevalence and concentrations of pathogen in food at consumption Account for growth and death dynamics of microbes Based on surveillance studies, simulation, modelling Investigate scenarios to determine effects of processing changes on risk
Hazard characterisation Dose–response assessment	NOEL derived from bioassay with safety factor calculate acceptable daily intake (ADI)	Data from human volunteer studies, animal models, outbreak investigations, used to estimate effect of varying levels of contamination Usually involves complex modelling
Risk characterisation	Should be negligible risk if regulatory compliance achieved	Risk estimate expressed in terms of probability of illness or death, e.g. from one portion of food, expected number of cases/100 000 population, etc., estimates for subgroups of population

(2) Microbiological risks are primarily the result of single exposures, whereas chemical risks are often due to cumulative effects.

(3) Veterinary drugs are approved for intentional administration to animals and approved uses can be structured to minimise human exposure to residues. Conversely, microbial contaminants are naturally occurring and exposure cannot be so readily manipulated.

(4) Micro-organisms are rarely homogenously distributed in food.

(5) Micro-organisms can be distributed via secondary transmission (i.e. person to person), in addition to direct ingestion of food.

(6) The exposed population may exhibit short-term or long-term immunity. This will vary according to the hazardous micro-organism.

On the international scene, the Uruguay Round Agreements (the SPS Agreement in particular) established the tenet that sanitary measures should be implemented on the basis of an assessment of the risk as appropriate to the circumstances (Section 1.7.1). To date, and with specific regard to microbiological food safety, much emphasis has been placed on the development of several frameworks for risk assessment (e.g. CAC 1998; CCFH 1998; ILSI 2000) and methodological advances. An increasing number of completed studies have been published or are in development under the aegis of the Joint FAO/WHO activities on risk assessment of microbiological hazards in foods. In response to the CAC, the FAO and WHO established a Joint Expert Meeting on Microbiological Risk Assessment (JEMRA) which is an expert consultative process to collect, collate and evaluate risk assessment data for significant pathogens in food at the international level. This process involves:

(1) Preparation of scientific descriptions of state-of-the-art knowledge

(2) Sharing this knowledge with all interested parties

(3) Interaction with risk managers to focus the scientific work towards areas where prevention is feasible

(4) Scientific scrutiny of the data presented

(5) Preparation of reports to enable an evaluation of the risk and an answer to the specific management question.

The CCFH is responsible for the risk management of food in international trade. Hence the JEMRA reports on risk assessment provide the scientific basis for the recommendations of the CAC committee on risk management.

Together the WHO and FAO have begun a number of major microbiological risk assessments. Following consultation with the CAC, who act as 'risk managers' for the international trade in food, a list of the most important food-borne, pathogenic micro-organisms was developed. From this list three food-borne pathogen–commodity combinations were chosen for initial risk assessment work:

(1)   *S.* Enteritidis in shell eggs and egg products (FSIS 1998)
(2)   *Salmonella* in poultry (Ebel *et al.* 2000; Kelly *et al.* 2000; Fazil *et al.* 2000a)
(3)   *Listeria monocytogenes* in ready-to-eat food (Buchanan & Lindqvist 2000; Ross *et al.* 2000)

This has now been completed and the assessments are covered in greater detail in Sections 4.2–4.4 (see Tables 1.10 and 3.2). Additionally they can be downloaded (pdf format) from the Internet (see Internet Directory). A number of other related publications have been released and many can be downloaded in pdf format from the WHO web site, including those of JEMRA. During 2001, JEMRA will initiate two new pathogen–commodity combinations: *C. jejuni* or *C. coli* in poultry and *Vibrio* spp. in seafood. It is proposed that JEMRA will continue to meet twice per year and evaluate further pathogen–commodity combinations. The WHO envisages that two to four pathogen–commodity combinations will be risk assessed to varying degrees each year. The WHO/FAO (2000a) preliminary report on hazard characterisation considered both food- and water-borne pathogens together because a significant number of 'food poisoning' outbreaks are caused by fruit and vegetables which have been irrigated with contaminated water. To date the majority of quantitative risk assessments have been for bacterial hazards rather than viruses, toxigenic fungi and parasitic protozoa because of the greater volume of data available.

Table 3.2   Joint FAO/WHO Expert Meetings on microbiological risk assessment (JEMRA).

The first meeting of this expert group was held 17–21 July 2000 and discussed preliminary reports on:

(1)   Hazard identification and hazard characterisation of *Listeria monocytogenes* in ready-to-eat foods (Buchanan & Lindqvist 2000)
(2)   Hazard identification and hazard characterisation of *Salmonella* in broilers and eggs (Fazil *et al.* 2000a). Note that this study had been withdrawn at the time of writing
(3)   Exposure assessment of *Salmonella* Enteritidis in eggs (Ebel *et al.* 2000)
(4)   Exposure assessment of *Salmonella* spp. in broilers (Kelly *et al.* 2000)
(5)   WHO/FAO guidelines on hazard characterisation for pathogens in food and water (JEMRA 2000).

The final documentation is expected October 2001. See Internet Directory (FAO/WHO) for web sites to update this topic.

Although it is accepted that the formalised use of risk analysis in food microbiology is in its infancy, it is likely that in the near future microbiological risk assessment will have a greater importance in the determination of the level of consumer protection that a government considers necessary and achievable.

3.2.1   Risk assessment

The goal of risk assessment is to provide an estimate of the level of illness from a pathogen in a given population. To achieve this the process must be scientifically based consisting of the following steps:

(1)   Statement of purpose
(2)   Hazard identification
(3)   Exposure assessment: the amount of hazard ingested
(4)   Hazard characterisation: the effect of the hazard and may include a dose–response assessment
(5)   Risk characterisation: the probability of illness and its severity
(6)   Production of a formal report.

Figure 3.2 shows the sequence of steps.

Risk assessment requires a multi-disciplinary approach including, for example, microbiologists, epidemiologists and statisticians. The terms 'quantitative risk assessment' and 'quantitative microbial risk assessment' are sometimes used. These emphasise the numerical expressions of risk and the related uncertainties; see Section 3.3 for more detail.

3.2.2   Risk management

Risk management is a process distinct from risk assessment. It entails comparing policy alternatives in consultation with all interested parties considering risk assessment and other factors relevant for the health protection of consumers and for the promotion of fair trade practices, and, if needed, selecting appropriate prevention and control options. Currently HACCP (Section 2.3) is a major means of hazard control. See Section 3.4 for more detail.

3.2.3   Risk communication

Risk communication is the interactive exchange, throughout the risk analysis process, of information and opinions concerning hazards and risks, risk-related factors and risk perceptions, among risk assessors, risk managers, consumers, industry, the academic community and other

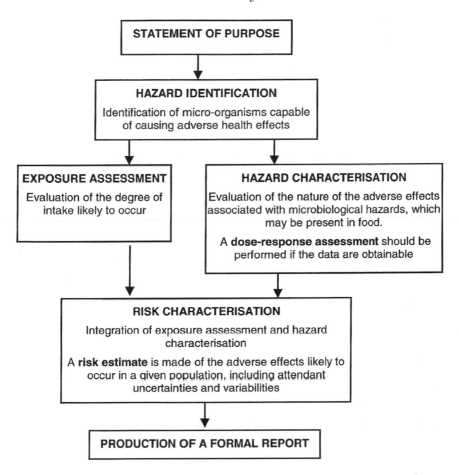

Fig. 3.2 Risk assessment flowchart (adapted from Notermans *et al.* (1996)).

interested parties, including the explanation of risk assessment findings and the basis of risk management decisions. See Section 3.5 for more detail.

3.3 Risk assessment

The goals of a risk assessment are to estimate the risk of illness in a given population from a pathogen and to understand the factors that influence it. Starting with a statement of purpose, the process (as defined by the *Codex Alimentarius*) is shown in Fig. 3.2. The definitions of these components are based on the Codex document (CAC 1999); there are eleven principles involved as given in Table 3.3.

Table 3.3   EU principles of microbiological risk assessment (CAC 1999).

Principle
(1) Microbiological risk assessment should be soundly based on science
(2) There should be a functional separation between risk assessment and risk management
(3) Microbiological risk assessment should be conducted according to a structured approach that includes hazard identification, exposure assessment, hazard characterisation and risk characterisation
(4) A microbiological risk assessment of microbiological hazards should clearly state the purpose of the exercise, including the form of risk estimate that will be the output
(5) The conduct of a microbiological risk assessment should be transparent
(6) Any constraints that impact on the risk assessment such as cost, resources or time, should be identified and their possible consequences described
(7) The risk estimate should contain a description of uncertainty and where the uncertainty arose during the risk assessment process
(8) Data should be such that uncertainty in the risk estimate can be determined: data and data collection systems should, as far as possible, be of sufficient quality and precision that uncertainty in the risk estimate is minimised
(9) A microbiological risk assessment should explicitly consider the dynamics of microbiological growth, survival and death in foods, and the complexity of the interaction (including sequelae) between human and agent following consumption as well as the potential for further spread
(10) Wherever possible, risk estimates should be reassessed over time by comparison with independent human illness data
(11) A microbiological risk assessment may need re-evaluation, as new relevant information becomes available

Risk assessment collates information on food-borne hazards that enables decision makers to identify interventions (risk mitigations) leading to improved public health. This may include regulatory action, voluntary activities and educational initiatives. Risk assessment also can be used to identify data gaps and target research that should have the greatest value in terms of public health impact.

The knowledge in each step of the risk assessment is combined to represent a cause-and-effect chain from the prevalence and concentration of the pathogen (exposure assessment) to the probability and magnitude of health effects (risk characterisation) (Lammerding & Paoli 1997). In risk assessment, 'risk' consists of both the probability and impact of disease. Therefore risk reduction can be achieved either by reducing the probability of disease or by reducing its severity.

3.3.1 Statement of purpose

The specific purpose of the risk assessment should be clearly stated and the output form, and possible output alternatives, should be defined. This stage refers to problem formulation and is intended to form a practical framework and a structured approach either for a full risk assessment or for a stand-alone process (such as hazard characterisation; Notermans & Teunis 1996; McNab 1998). During this stage, the cause of concern, the goals, breadth and focus of the risk assessment should be defined. The statement may also include data requirements, as they may vary depending on the focus and the use of the risk assessment and the questions relating to uncertainties that need resolving. Output might, for example, take the form of a risk estimate of an annual occurrence of illness, or an estimate of annual rate of illness per 100 000 population, or an estimate of the rate of human illness per eating occurrence.

3.3.2 Hazard identification

Definition: Hazard identification consists of the identification of biological, chemical and physical agents (micro-organisms and toxins) capable of causing adverse health effects which may be present in a particular food or group of foods.

Hazard identification consists of the identification of biological, chemical and physical agents (micro-organisms and toxins) capable of causing adverse health effects, which may be present in a particular food or group of foods. It often involves evaluating epidemiological data linking foods and pathogens to human illness. Hazard identification can be used as a screening process to identify pathogen-commodity combinations of greatest concern to the risk managers (Lammerding & Fazil 2000). Information on potentially hazardous micro-organisms and toxins can be obtained from numerous sources, such as government surveillance studies and various highly reputable organisations (e.g. ICMSF publications). The information may describe microbial growth and death conditions (pH, a_w, D values; Section 2.5). Hazard identification is easier in microbial risk assessments than chemical risk assessments because the microbiological hazards are well recognised due to their short incubation period (days), whereas in chemical risk assessment the adverse health effect may require a long time-period (years) after exposure to become apparent (see Table 3.1). Micro-organisms associated with food-borne illness are given in Tables 1.1 and 1.2.

The key to hazard identification is the availability of public health data and a preliminary estimate of the sources, frequency and amount of the hazard(s) under consideration. Although food-borne bacterial pathogens with certain foods are well recognised, surveillance data and epidemiological studies can reveal high-risk products and processes. The information collected is subsequently used in 'exposure assessment' where the effect of food processing, storage and distribution (covering from processing to consumption) on the number of food-borne pathogens is assessed.

3.3.3 Exposure assessment

Definition: Exposure assessment is the qualitative and/or quantitative evaluation of the likely intake of biological, chemical and physical agents via food as well as exposure from other sources if relevant.

Exposure assessment determines the likelihood of consumption and the likely dose of the pathogen to which the consumers may be exposed in a food. The assessment should be in reference to a specified portion size of food at the time of consumption or a specified volume of water consumed per day. Overall, it describes the pathways through which a hazardous micro-organism enters the food chain and is subsequently distributed and challenged in the production, distribution and consumption of food. This may include an assessment of actual or anticipated human exposure. For food-borne microbiological hazards, exposure assessment might be based on the possible extent of food contamination by a particular hazard, and on consumption patterns and habits. Exposure to food-borne pathogens is a function of the frequency and amount of food consumed, and the frequency and level of contamination.

The steps in food production that affect human exposure to the target organism from primary production to consumption are described as the 'farm-to-fork' sequence (also known as the process risk model, Section 4.5.1). It is convenient in exposure assessment to divide the sequence into a series of modules as shown in Fig. 3.3. The diagram emphasises the two sets of data required in a quantitative risk assessment: prevalence and concentration of the specified pathogen (hazard). Depending upon the scope of the risk assessment, exposure assessment can begin with either the pathogen prevalence in raw materials or with the description of the pathogen population at subsequent steps, such as during processing. Where surveillance data is lacking or insufficient, the effect of processing on prevalence and concentration can be modelled using predictive microbiology (Section 2.7). Nevertheless, such predicted values should be verified with surveillance data where possible.

A flow diagram for exposure assessment is given in Fig. 3.4. This should be combined with the activities described for hazard characterisation in Fig. 3.5.

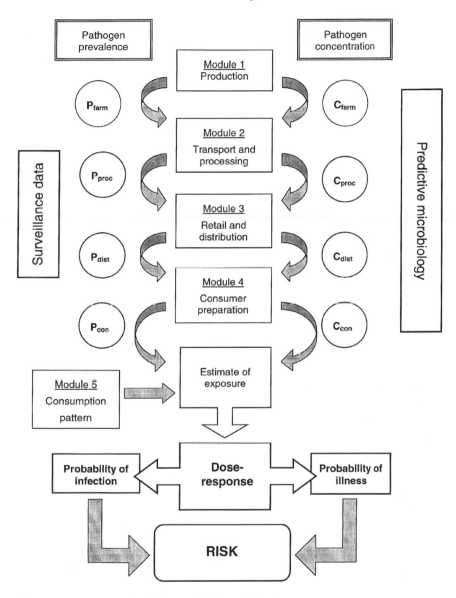

Fig. 3.3 Framework of 'farm-to-fork' modules for exposure assessment.

Exposure assessment is one of the most complex and uncertain aspects of microbial risk assessment. Hence, modelling and simulation studies are required (Section 2.7). Great emphasis must be placed on estimating the effects of a large number of factors on the microbial population. These factors include the following:

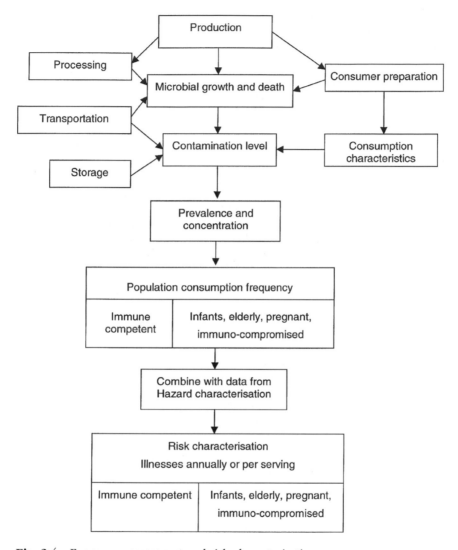

Fig. 3.4 Exposure assessment and risk characterisation.

- The microbial ecology of the food
- Microbial growth requirements (intrinsic, extrinsic parameters; Table 2.3)
- The initial contamination of the raw materials
- Prevalence of infection in food animals
- The effect of the production, processing, cooking, handling, storing, distribution steps and preparation by the final consumer on the microbial agent (i.e. the impact of each step on the level of the pathogenic agent of concern)

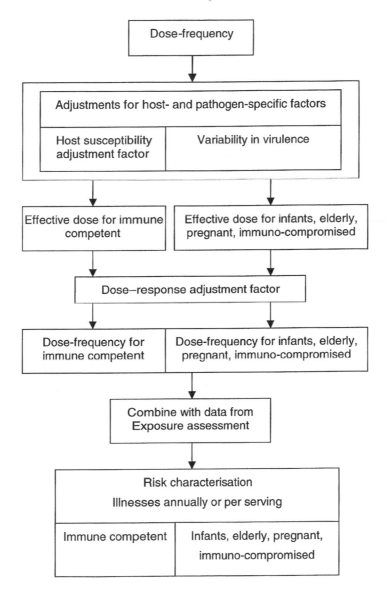

Fig. 3.5 Hazard characterisation and risk characterisation.

- The variability in processes involved and the level of process control
- The level of sanitation, slaughter practices, rates of animal–animal transmission
- The potential for (re-)contamination (e.g. cross contamination from other foods and re-contamination after a heat treatment)

- The methods or conditions of packaging, distribution and storage of the food (e.g. temperature of storage, relative humidity of the environment, gaseous composition of the atmosphere), the characteristics of the food that may influence the potential for growth of the pathogen (and/or toxin production) in the food under various conditions, including abuse (e.g. pH, moisture content or water activity, nutrient content, presence of antimicrobial substances, competitive flora).

Since preformed microbial toxins and viruses do not grow in food, their exposure assessment is simpler than that of bacteria which may multiply, die or adapt during the farm-to-fork stages.

The factors related to the food matrix are principally those that may influence the survival of the pathogen through the hostile environment of the stomach. They may include:

- Composition and structure of the food matrix e.g. highly buffered foods
- Entrapment of bacteria in lipid droplets
- Processing conditions (e.g. increased acid tolerance of bacteria following pre-exposure to moderately acid conditions)
- Conditions of ingestion (e.g. initial rapid transit of liquids through an empty stomach).

Data on microbial survival and growth in foods can be obtained from food poisoning outbreaks, storage tests, historical performance data of a food process, microbiological challenge tests and predictive microbiology (Section 2.7). These tests provide information on the probable numbers of organisms (or quantity of toxin) present in a food at the point of consumption.

Information on consumption patterns and habits may include the following:

- Socio-economic and cultural background, ethnicity
- Consumer preferences and behaviour, because they influence the choice and the amount of the food intake (e.g. frequent consumption of high-risk foods)
- Average serving size and distribution of sizes
- Amount of food consumed over a year, considering seasonality and regional differences
- Food preparation practices (e.g. cooking habits and/or cooking time, temperature used, extent of home storage and conditions, including abuse)
- Demographics and size of exposed population(s) (e.g. age distribution, susceptible groups).

Bettcher *et al.* (2000) and Ruthven (2000) published surveys of food consumption for various regions of the world. Consumption values varied; for example, the consumption of chicken meat ranged from 11.5 g (Far Eastern diet) to 44.0 g (European diet).

3.3.4 Hazard characterisation

Definition: Hazard characterisation is the qualitative and/or quantitative evaluation of the nature of the adverse effects associated with biological, chemical and physical agents that may be present in food. If data are available, then a dose–response assessment should be performed.

Hazard characterisation provides an estimate of the nature, severity and duration of the adverse effects following ingestion of the hazard, i.e. for a given number of micro-organisms consumed at a sitting, what is the probability of illness? If sufficient data are available, then a dose–response relationship is determined (see below). This step, like exposure assessment, is very complex. Factors important to consider relate to the micro-organism, the food and the host. A flow diagram for hazard characterisation is given in Fig. 3.5; this should be combined with the activities described for exposure assessment in Fig. 3.4. A preliminary guideline document on hazard characterisation has been released by WHO/FAO (2000a); see Internet Directory for URL.

Hazard characterisation can be a stand-alone process as well as a component of risk assessment. The hazard characterisation must be transparent (assumptions and variables well documented) such that risk managers can combine the information with an appropriate exposure assessment. This could even occur by combining the two components across different countries. A hazard characterisation developed for water exposure may be adapted to a food exposure scenario with modification for the food matrix effects. In general, hazard characterisations are fairly adaptable between risk assessments for the same pathogen. This contrasts with exposure assessments, since those are highly specific to the production, processing and consumption patterns within a country or region.

Ingestion of a pathogen does not necessarily mean the person will become infected, nor that illness or death will occur. As shown in Fig. 3.6 there are a number of barriers to infection and illness. These barriers can be compromised as the result of host and food matrix factors. The response (infection, illness, death) to pathogen ingestion will vary according to pathogen, food and host factors; this is commonly known as the 'infectious disease triangle' (Fig. 3.7).

Pathogen factors
In determining the hazard characterisation for food-borne microbial

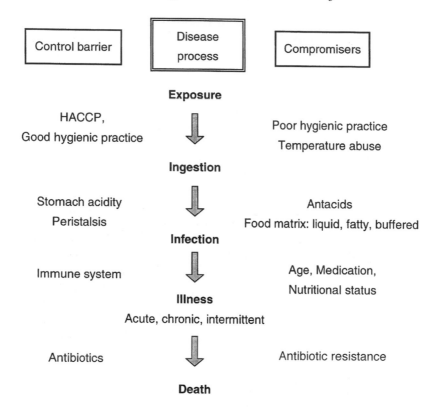

Fig. 3.6   Barriers to infectious diseases.

pathogens, the biological aspects of the pathogen must be considered. Infection can be seen as resulting from the successful passage of multiple barriers in the host (Fig. 3.6). These barriers are not equally effective in eliminating or inactivating pathogens. Each individual pathogen has some particular probability or relative frequency to overcome a barrier, which is conditional on the previous step(s) being completed successfully.

The infectivity of microbial pathogens is dependent upon many factors. These can be intrinsic to the pathogen (phenotypic and genetic characteristics) as well as host specificity. The stress response to temperature, drying, acidity, etc., may affect the virulence of the pathogen (Section 2.6). Figure 3.7 lists the various pathogen factors which must be considered in hazard characterisation.

Food-related factors
The food matrix can affect the survival of the microbial pathogen. For example, a food with a high fat content can protect the organism from stomach acid, and hence increase the chances that it will survive and

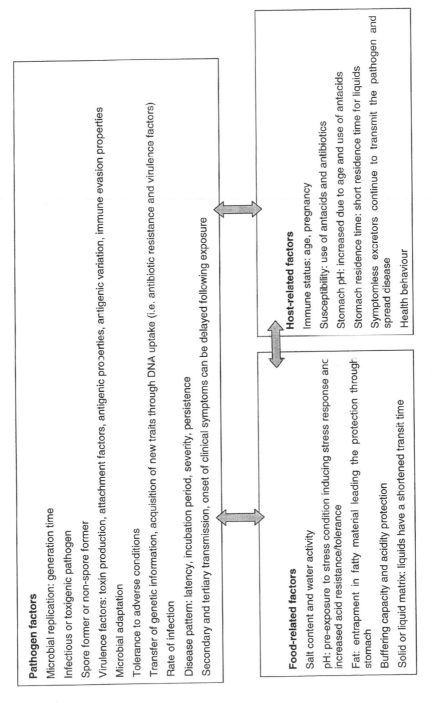

Pathogen factors

Microbial replication: generation time

Infectious or toxigenic pathogen

Spore former or non-spore former

Virulence factors: toxin production, attachment factors, antigenic properties, antigenic variation, immune evasion properties

Microbial adaptation

Tolerance to adverse conditions

Transfer of genetic information, acquisition of new traits through DNA uptake (i.e. antibiotic resistance and virulence factors)

Rate of infection

Disease pattern: latency, incubation period, severity, persistence

Secondary and tertiary transmission, onset of clinical symptoms can be delayed following exposure

Host-related factors

Immune status: age, pregnancy

Susceptibility: use of antacids and antibiotics

Stomach pH: increased due to age and use of antacids

Stomach residence time: short residence time for liquids

Symptomless excretors continue to transmit the pathogen and spread disease

Health behaviour

Food-related factors

Salt content and water activity

pH: pre-exposure to stress condition inducing stress response and increased acid resistance/tolerance

Fat: entrapment in fatty material leading the protection through stomach

Buffering capacity and acidity protection

Solid or liquid matrix: liquids have a shortened transit time

Fig. 3.7 Factors affecting the infectivity of microbial pathogens.

cause infection. The food processing can give the organism a heat-shock which may cause microbial adaptation, including acid tolerance, and hence increased survival through the stomach (Section 2.5). Exposure to short chain fatty acids can induce acid resistance in *Salmonella* serovars. Additionally, if the food is highly proteinaceous it may act as a buffer and protect the pathogen from stomach acid.

Host-related factors

As shown in Fig. 3.7, there are a variety of host-related factors that influence the disease response. The well recognised factors are age and immune status. It has been estimated that 20% of the total population may be immunoimpaired and hence are separately described in Fig. 3.6 (Gerba *et al.* 1996). Prior exposure is of limited importance for food-borne pathogens since many do not invade the host body; however, it may be important for the protozoal parasite *Cyclospora cayetanensis*. See Section 4.2.2 for a description of human feeding trials with *Salmonella* serovars.

Hazard characterisation must consider the range of biological responses to pathogen ingestion. This ranges from symptomatic infections to illness (acute, chronic and intermittent) and death (see Fig. 3.6). The severity of the risk may be expressed as duration of the illness, proportion of the population affected or as mortality rate, and should identify the 'at-risk' groups. For pathogens that cause chronic sequelae (e.g. Guillain–Barré syndrome, Section 1.2), the effect on quality of life may be included in the hazard characterisation.

Figures 3.6 and 3.7 show the various host-related factors that affect susceptibility to microbial infections and severity of illness. The important aspect of hazard characterisation is to provide information on who is at risk and the associated severity for the susceptible subpopulations. In Section 1.6 the cost of food-borne disease in the USA was given as between US$6.6 and 37.1 billion. The economic and social cost is so extensive that the WHO has identified improved food safety as one of its aims for the twenty-first century (Sections 1.1 and 1.8).

3.3.5 Dose–response assessment

The goal of a dose–response assessment is to determine the relationship between the magnitude of exposure (dose) to the pathogen and the severity and/or frequency of adverse health effects (response). Sources of information include:

(1) Human volunteer studies
(2) Population health statistics

(3) Outbreak data
(4) Animal trials.

Because of the variety of disease response (Fig. 3.6) the end-point must be clearly delinated. Essentially there are four possible responses to a dose:

(1) Probability of infection following ingestion
(2) Probability of illness (morbidity) following infection
(3) Probability of chronic sequelae following illness
(4) Probability of death (mortality).

It is generally assumed that the effects of food-borne pathogens are dose-dependent but not cumulative (unlike many chemical hazards). Hence the frequency of consumption must be determined because multiple exposures to low doses may not represent the same risk as a single exposure to a large dose.

The dose–response relationship is complex and in many cases may not be demonstrable. For example, Medema *et al.* (1996) reported that although the rate of *C. jejuni* infection was dose related, the rate of illness was not. This contrasts with *Salmonella* infections where higher doses are reported to result in greater frequency of severe illness (Coleman & Marks 1998).

Currently there is a lack of data concerning pathogen-specific responses, the effect of the host's immunocompetence on the pathogen-specific responses, translation of infection into illness and of illness into different outcomes. Also, as given in Fig. 3.7, there are numerous variable factors involved, such as

- Physiology, virulence and pathogenicity of the micro-organisms
- Variation in host susceptibility
- Food matrix.

It is therefore essential that the dose–response analysis clearly identifies which information has been used and its source. In addition, variability (due to known factors such as amount of food consumed and population susceptibility) and uncertainty (insufficient experimental data or lack of knowledge of the pathogen/host/food being studied) in the data should be thoroughly described for the risk assessment to be transparent (Nauta 2000).

An important issue is whether a threshold, or a collaborative action of the pathogens, may be a plausible mechanism for any harmful effect, or whether a single micro-organism may cause adverse effects under certain circumstances. When extrapolating from laboratory animal or *in vitro* studies, the information on the biological mechanisms is important with respect to assessment of relevance to humans. Aspects to be considered are outlined in Table 3.4.

Table 3.4   Aspects of the dose–response relationship (WHO/FAO 2000a).

1   Organism type and strain
2   Route of exposure
3   Level of exposure (the dose)
4   Adverse effect considered (the response)
5   Characteristics of the exposed population
6   Duration – multiplicity of exposure

Until relatively recently it had been assumed that there was a threshold level of pathogens that must be ingested for the micro-organism to produce an infection or disease (the minimum infectious dose; see Table 3.5). This approach has been largely superceded by the proposal that infection may result from the survival of a single, viable, infectious pathogenic organism ('single-hit concept'). Thus, regardless of how low the dose, there is always a non-zero effect of infection and illness. It should be noted that the accuracy of the infectious dose is debatable. For example, it is commonly cited that the infectious dose for *C. jejuni* is as low as 500 bacteria. However, this value can be traced back to the paper by Robinson (1981) who performed the investigation on himself. After a light breakfast he drank a glass of milk that contained 500 bacteria, which consequently made him ill. In contrast, Martin *et al.* (1995) gave the probability of developing at least light symptoms as 24% when consuming at the level of 10^2 *C. jejuni* cells but as 32% when ingesting 10^8 *C. jejuni* cells. The 'tolerable' level for *B. cereus* is accepted to be less than 10^4 cells (Section 4.6).

Concurrently with the analysis of raw clinical or epidemiological information or data, mathematical modelling has been advocated to provide assistance in developing dose–response relationships, in particular when extrapolation to low doses is necessary. An active area of research is the development of more appropriate mathematical models for dose–response assessment.

3.3.6   Dose–response models

Food is frequently contaminated with smaller numbers of microbial pathogens than those used in laboratory trials (human feeding studies and animal models). Therefore mathematical models are needed to extrapolate low dose responses from the high dose data. Various dose–

Table 3.5 Minimum infectious dose (Forsythe 2000).

Organism	Estimated infectious dose
Non-spore-forming bacteria	
C. jejuni	1000
Salmonella spp.	10^4-10^{10}
Sh. flexneri	10^2-10^9
Sh. dysenteriae	10-10^4
E. coli	10^6-10^7
E. coli O157:H7	10-100
St. aureus	10^5-$<10^6$/g[a], 0.5-5 µg toxin
V. cholerae	1000
V. parahaemolyticus	10^6-10^9
Y. enterocolitica	10^7
Spore-forming bacteria	
B. cereus	10^4-10^8
Cl. perfringens	10^3-10^{5a}
Cl. botulinum	10^6-10^7, 0.5-5 ng toxin
Viruses	
Hepatitis A	< 10 particles
Norwalk-like virus	< 10 particles
Protozoa	
Cryp. parvum	10 oocysts
Entamoeba coli	1 cyst

[a] Viable count able to produce sufficient toxin to elicit a physiological response.

response models have been proposed to describe the relation between ingestion of a certain number (*N*) of a pathogenic micro-organism and the possible outcomes. The main models are exponential and beta-Poisson (Holcomb *et al.* 1999).

One approach assumes that each micro-organism has an inherent minimal infective dose, i.e. there is a threshold value below which no response (depending upon the end-point) is seen. The value of the minimal dose in the population may be assumed to follow different distributions. The alternative approach is that the actions of individual cells of pathogenic micro-organisms are independent and that a single micro-organism has the potential to infect and provoke a response in the individual, i.e. a single-hit, non-threshold model (Haas 1983). The exponential model assumes that the probability of a single cell causing infection is independent of dose. In contrast, the beta-Poisson model assumes that infectivity is dose-dependent. Different microbial pathogens appear to fit different dose–response models.

Data for protozoan parasites can be well described by the exponential models.

Exponential model

$$P_i = 1 - \exp(-r * N)$$

where P_i is the probability of infection, r is host/micro-organism interaction probability, and N is the ingested dose of micro-organisms. This has been used for *Cryptosporidium parvum* and *Giardia lamblia* (Teunis *et al.* 1996), though Holcomb *et al.* (1999) modified it slightly into the Simple exponential model and Flexible exponential model (see Rose *et al.* 1991).

Simple exponential model

$$P_i = 1 - \exp(-r * \log_{10}N)$$

Flexible exponential model

$$P_i = \beta * [1 - p * \exp(-\varepsilon\{\log_{10}N - x\})]$$

where P_i is the probability of infection, r is the host/micro-organism interaction probability, N is the ingested dose of micro-organisms, β is the asymptotic value of probability of infection as the dose approaches $\beta = 1$ (Holcomb *et al.* 1999), x is the predicted dose at a specified value of p where $p = 1 - Pr^*$, and ε is the curve rate value affecting the spread of the curve along the dose axis.

In contrast, the bacterial infection data are generally well described using beta-Poisson models (Haas 1983; Teunis *et al.* 1997, 1999) and the Weibull-gamma model (Todd & Harwig 1996; Holcomb *et al.* 1999).

Beta-Poisson model

$$P_i = [1 - (1 + N/\beta)]^{-\alpha}$$

where P_i is the probability of infection, N is the ingested dose of micro-organisms, and α and β are parameters that are specific to the pathogen affecting the shape of the curve (see Vose 1998).

The beta-Possion model is frequently used for describing dose–response relationships when assessing low levels of bacterial pathogens. It generates a sigmoidal dose–response relationship that assumes no threshold value for infection (see Figs 3.8 and 3.9 for dose response curves for *Salmonella* and *C. jejuni*, respectively). Instead it assumes that there is a small but finite risk that an individual can become infected

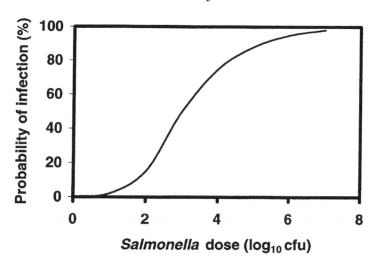

Fig. 3.8 Beta-Poisson dose–response curve for *Salmonella* serovars.

after exposure to a single cell of a bacterial pathogen (single-hit concept).

Marks *et al.* (1998) compared two beta-Poisson models for a risk assessment of *E. coli* O157:H7 in hamburgers, one of which had a threshold value of three bacteria. The difference between the models was only significant in the low dose range, and the resulting estimates of risk

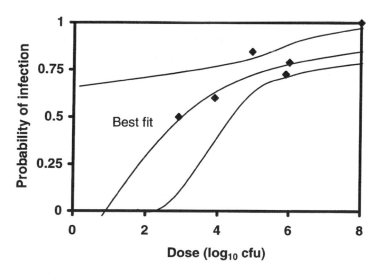

Fig. 3.9 *C. jejuni* dose–response in human feeding trials (Black *et al.* 1988; Fazil *et al.* 2000b).

were 100- to 1000-fold larger using the non-threshold model, depending upon the cooking temperature. They concluded that the two-parameter beta-Poisson model seemed inadequate as a default model for describing the complexity of dose–response interactions, especially in cooked foods.

Weibull-gamma model

$$P_i = 1 - [1 + (N)^b/\beta]^{-\alpha}$$

where P_i is the probability of infection, N is the ingested dose of micro-organisms, and α, β and b are parameters that are specific to the pathogen affecting the shape of the curve.

The Weibull-gamma model assumes that the probability that any individual cell can cause an infection is distributed as a gamma function. Hence this model is very flexible in that its shape depends on the parameters selected and has more recently been used for dose–response analyses (Farber *et al.* 1996).

Beta-binomial model

Cassin *et al.* (1998a) developed a beta-binomial dose–response model for *E. coli* O157:H7 in hamburgers which yields variability for probability of illness from a particular dose, in contrast to the original model which only specifies a mean population risk (Section 4.5.1).

$$P = 1 - (1 - P_i(1))^N$$

where $P_i(1)$ is the probability of illness from ingestion of one micro-organism, and this probability was assumed to be beta-distributed with parameters α and β.

By fitting the model to data from human feeding studies with *Sh. dysenteria* and *Sh. flexneri* (Crockett *et al.* 1996) a dose–response curve was generated which showed the estimated uncertainty in the average probability of illness versus the ingested dose. The variability between feeding studies was used for the uncertainty in the parameters α and β.

The same model may not be equally effective for all biological endpoints caused by the pathogen. For example, the exponential model did not fit data for *L. monocytogenes* infection of mice (isolation from spleen and liver), but was good for describing the relationship between the dose and the frequency of death (FDA 1999; FAO/WHO initiative on microbial risk assessment, see Internet Directory). The Gompertz equation gave the best fit for frequency of infection (Coleman & Marks 1998).

Gompertz model

$$P_i = 1 - \exp[-\exp(a + bf(\times))]$$

where *a* is a model (intercept) parameter, *b* is a model (slope) parameter and *f(n)* is a function of dose.

It is important to note that unit dose and biological dose are almost always different as a result of the non-homogenous distribution of micro-organisms in food (see Table 3.6). A range of α and β values is given in Table 3.7. Infection values are different from the minimum infectious dose values commonly given (see Table 3.5).

Table 3.6 Definitions of 'dose'.

Measured dose	The dose as estimated by the mean concentration in the delivery matrix. Note that the dose unit (e.g. colony forming units) may contain one or more discrete organisms
Functional dose	The measured dose corrected for the sensitivity and specificity of the measurement method to detect viable, infectious agents
Administered dose	The number of viable, infectious agents actually administered, whether orally (ingestion, gavage, nasal–gastric intubation), by injection (intraperitoneal or intravenous), or by other means. Although this is the dose that is actually given to an individual, it can only be estimated
Effective dose	The number of viable, infectious agents that actually reach the site of infection

Other models
It is probable that alternative dose–response models will be required according to whether the microbial pathogen is infectious or toxigenic, and whether the organism produces the toxin whilst passing through the intestinal tract (*Cl. perfringens*) or preformed toxin is ingested in the food (*St. aureus*). Dose-response modelling is further complicated by micro-organisms such as *B. cereus* which cause two different illnesses, emetic and diarrhoeal (Section 4.6). Some organisms may tend to remain as colonies or clumps following ingestion, and hence infection due to a single cell may not frequently occur. Virulence mechanisms, such as attachment and effacement of enterocytes as found in pathogenic *E. coli*, may make the intestinal wall more susceptible to further infection and illness, in contrast to the biological response to toxins (e.g. *St. aureus*

Table 3.7 Dose–response parameters for food- and water-borne pathogens, where α and β are beta-Poisson parameters and N_{50} represents the ID_{50}[a].

Micro-organism	Model	Model parameters	Reference
Non-typhi *Salmonella*	beta-Poisson	$\alpha = 0.4059$ $\beta = 5308$	Fazil *et al.* (2000a)
E. coli	beta-Poisson	$\alpha = 0.1705$ $\beta = 1.61 \times 10^6$	Rose *et al.* (1995)
Echovirus 12	beta-Poisson	$\alpha = 0.374$ $\beta = 186.69$	Schiff *et al.* (1984)
Rotavirus	beta-Poisson	$\alpha = 0.26$ $\beta = 0.42$ $\alpha = 0.265$ $N_{50} = 0.42$	Ward *et al.* (1986) Gale (2001) Hauschild *et al.* (1982)
Cryptosporidium	Exponential	$r = 0.004191$	Medema & Schijven (2001)
Giardia lamblia	Exponential	$r = 0.02$	Medema & Schijven (2001)

[a] ID_{50} is the dose causing 50% infection.

enterotoxins and aflatoxins) which are adsorbed through the intestinal wall without damaging it. For an extensive treatment on the mathematical modelling of dose–response relationships and data-fitting the reader is referred to Haas *et al.* (1999).

3.3.7   Dose and infection

In general, dose–response models for food-borne microbial pathogens should consider the discrete (particulate) nature of organisms and should be based on the concept of infection from one or more 'survivors' from an initial dose. There are, however, different definitions of dose, and hence this must be clearly stated in any study (see Table 3.6). The measured dose may need to be corrected according to the sensitivity and/or specificity of the detection method which may not be specific for the viable, infectious organism. Therefore, the functional dose is defined as the corrected measured dose. It should be recognised that the standard agar plate viability count gives 'colony forming units per gram'; however, a colony may have grown from one or more initial bacterial cells. Hence the method will tend to underestimate the number of infectious bacterial cells. An additional complication in estimating the dose is the non-random distribution of micro-organisms in food.

The functional dose describes the average number of viable, infectious units in the inoculum. Every individual in a population will consume a subsample containing a discrete, but unknown, number of units. This ingested dose can be characterised by a frequency distribution such as a Poisson distribution for random events.

Each individual organism in the ingested dose is assumed to have a distinct probability of surviving all barriers to reach a target site for colonisation. The relation between the actual number of surviving organisms (the effective dose) and the probability of colonisation of the host is a key concept in the derivation of dose–response models.

Infection is most commonly defined as a situation in which the pathogen, after ingestion and surviving all barriers (Fig. 3.6), actively grows at its target site. It can be measured by different methods, such as faecal excretion or immunological response. Hence apparent infection rates may differ from actual infection rates, depending upon the sensitivity and specificity of the diagnostic assays. Target sites may be specific (one type) or non-specific (many types), and local (non-invasive) or systemic (invasive). The sequence of events and time required for each event to occur may be important and may vary by pathogen.

Infections may be asymptomatic (i.e. the host does not develop any adverse reactions to the infection and clears the pathogens within a limited period of time). The probability of sequelae and/or mortality for a given illness depends on the characteristics of the pathogen, but more importantly on the characteristics of the host. These are usually rare events that affect specific subpopulations. These may be identified by factors such as age or immune status but, increasingly, genetic factors are being recognised as important determinants.

The most obvious means for acquiring information on dose–response relations for food-borne microbial pathogens is to expose humans to the disease agent under controlled conditions. There have been a limited number of human feedings trials using volunteers, and most of these have been in conjunction with vaccine trials. From the human volunteer studies, Probability of infection (P_i) values have been determined for a number of food- and water-borne pathogens (Table 3.8); for example, there is a 1 in 2000 chance of an individual becoming infected from a single *Salmonella* cell compared to a 1 in 7 million chance from *V. cholerae*. Medema *et al.* (1996) used the human feeding trial data of Black *et al.* (1988) to determine the dose–response relationship with a beta-Poisson model (Fig. 3.9, Table 3.9) for infection with *C. jejuni*. The occurrence of symptoms did not follow a similar dose-related trend, however. The beta-Poisson model for *E. coli* O157:H7 (Section 4.5.1) showed considerable variability. Since no comparable data was available for *E. coli* O 157:H7,

Table 3.8 Probability of infection for food- and water-borne pathogens (from Bennett *et al.* 1987; Notermans & Mead 1996).

Enteric pathogen	Probability of infection (P_i)	Fatality/case (%)
Campylobacter jejuni	7×10^{-3}	0.1
Salmonella serovars	2×10^{-3}	0.1
Shigella spp.	1×10^{-3}	0.2
V. cholerae (classical)	7×10^{-6}	1.0
V. cholerae El Tor	1.5×10^{-5}	4.0
Rotavirus	3×10^{-1}	0.01
Giardia spp.	2×10^{-2}	ND[a]

[a] Not determined.

this study used the α and β values for *Sh. dysenteriae* due to the similarity in virulence.

There are a number of limitations associated with the use of human feeding trials:

(1) The main problem is that they are almost always conducted with healthy, young (18–50 years old) adults, usually men, whereas the most vulnerable members of society are the elderly, pregnant and the very young.
(2) Ethically, pathogens that are life threatening (such as *E. coli* O 157) or that cause disease only in high risk subpopulations (such as *L. monocytogenes* serotype 4b) are not amenable to such volunteer studies.
(3) Human feed trials usually only use a small number of volunteers per dose and a small number of doses. The average is six volunteers per dose, though the range is from 4 to 193 (see Table 3.9). Because the studies are often for vaccine trials, the dose ranges used are generally high to ensure a response in a significant portion of the test popu-

Table 3.9 Human trial data for *C. jejuni* infection (Black *et al.* 1988; Medema *et al.* 1996) (See also Fig. 3.9.)

Dose (cfu)	Number of volunteers	Number infected	Number showing symptoms
8×10^{2}	10	5	1
8×10^{3}	10	6	1
9×10^{4}	12	11	6
8×10^{5}	11	8	1
1×10^{6}	19	15	2
1×10^{8}	5	5	0

lation. Therefore often the doses are not in the region of most interest to dose-response modellers.

(4) The pathogen is often fed to the volunteer in a non-food matrix after neutralising the stomach acidity. Hence the size of dose reaching the intestines is probably different to normal ingestion patterns (Kothary & Babu 2001).

However, obtaining human volunteer data does have an advantage that interspecies conversions, as required from animal models, are not necessary.

Animal models depend on the selection of appropriate animal(s) showing the same disease response and a conversion factor to human response. The major advantage is that a larger number of replicates and dose range can be used, and the animals can be kept under more environmentally controlled conditions than human volunteers. Aside from ethical issues on animal experimentation, the animal models do have inherent limitations. The animals are often similar in age, weight and immune status. Hence, similar criticisms can be used as have been applied to human feeding trials. Additionally, laboratory animals have very limited genetic variation, unlike humans. The use of animal models for *L. monocytogenes* risk assessment is discussed in Section 4.4.1.

Many national governments and several international organisations compile health statistics for infectious diseases, including those that are transmitted by food and water. Such data are crucial for the characterisation of microbiological hazards. In addition, surveillance-based data have been used in conjunction with food survey data to estimate dose-response relations (Notemans *et al.* 1998b). Epidemiological studies are important as a means of verifying dose-response models. The effectiveness of dose-response models is typically assessed by combining them with exposure estimates and determining if they approximate the annual disease statistics for the hazard.

Studies of outbreaks of food-borne illness may be very useful to dose-response modellers (Mintz *et al.* 1994). In two outbreaks of salmonellosis investigated by the Minnesota Department of Health, USA, the highest levels of *Salmonella* contamination detected in implicated vehicles were 4.3 organisms per 100 g of cheese and 6.0 organisms per 65 g (one-half cup) serving of ice cream. Previous outbreak investigations elsewhere in the United States and Canada have documented similar low level *Salmonella* contamination of implicated products such as cheese and chocolate. Estimates of infective doses of *Salmonella* provided by these outbreak investigations are several logarithms lower than estimates of minimum infective doses provided by clinical trials with a limited number of volunteers. Data from epidemiological investigations of outbreaks may not always be totally reliable because the information was not collected according to a standardised format or procedure. Estimates of attack rate may be overestimated or underestimated because they may be based on symptoms rather than laboratory-confirmed cases where the causative organism was recovered. According to WHO/FAO (2000) determination of the exposure dose in outbreak scenarios may be inaccurate because:

(1) Representative samples of the contaminated food or water were not obtained
(2) Detection methods may not be sufficiently accurate (e.g. *Cryptosporidium* oocysts in water)
(3) Estimates of water or food consumption levels are inaccurate.

Section 4.2.2 covers the recent extensive JEMRA report on hazard identification and hazard characterisation of *Salmonella* in broilers and eggs, and compares the dose–response models for feeding trials and outbreaks.

Buchanan *et al.* (2000) encouraged a more mechanistic approach to dose–response modelling that takes into account the limitations of extrapolating from human feeding trials of healthy (male) individuals to the general, diverse population. They proposed that there were three stages to be considered:

(1) Gastric acidity barrier
(2) Attachment/infectivity
(3) Morbidity/mortality.

Gastric acidity barrier
The infectivity of ingested pathogens depends initially upon their survival through the stomach contents. This depends upon the kill effect of the stomach acidity and the rate of gastric emptying. The rate of death can be expressed as a D value (Section 2.5.4) according to the equation.

$$\log_{10}D = (0.554 \times pH) - 1.429$$

The rate of emptying is given by

$$R(\%) = 100.4 \text{ min} + (-0.429 \times t)$$

where R (%) is the percentage retention and t is the time (min).

A combination of these equations predicts that at pH 2.2 (normal pH of gastric juice) only 1–2 cells per 100 will survive, whereas at pH 4.0 there would be approximately 50% survival. Buchanan *et al.* (2000) found these predicted results correlated with values derived experimentally by Peterson *et al.* (1989).

Attachment/infectivity

The second stage in the infection model is the ability of the ingested pathogen to overcome the wash-out effect due to the (relative to microbial size) rapid flow of intestinal contents. Hence the pathogen must attach to and colonise the intestinal epithelium. The attachment site varies according to the pathogen (i.e. microvilli via mannose-specific type 1 fimbrae in *Salmonella* serovars). Because of the lack of appropriate data, Buchanan *et al.* (2000) used a value that 1 in 100 surviving pathogens were able to attach and colonise the epithelial layer. Obviously further research is required on this topic.

Morbidity/mortality

The final stage in the infection model is the likelihood that the ingested pathogen will cause illness symptoms and even death. The progress of the infection will depend upon the virulence mechanisms of the pathogen and the host defence (primarily immune) system. The immune system deteriorates with age, and the defence system may be weakened due to inadequate diet and the use of prescription drugs such as antacids.

Hence the mechanistic model implies that the rate of infection is dependent upon the stomach pH, intestinal attachment, virulence mechanisms and host immune status. Buchanan *et al.* (2000) used the model to simulate the effect of *Salmonella* ingestion on two populations: elderly (> 65 years old) of which 30% suffered from achlorhydria (reduced gastric acid secretion) and adults between 20 and 65 years old, of which only 1% suffered from achlorhydria. The population sizes were set at 100 000 and the ingested dose was 100 *Salmonella* cells. Those with achlorhydria had stomach pH values of 4.0 compared with the normal pH 2.2 (Forsythe *et al.* 1988).

Based on a series of assumptions, such as rates of morbidity and mortality and proportion of each population becoming infected and dying, Table 3.10 was constructed (summarised from Buchanan *et al.* 2000). The model predicts the considerably greater number of deaths for the elderly. In this example the mechanistic model has only been applied to one dose, but could be further utilised for a range of exposure levels.

3.3.8 Risk characterisation

Definition: Risk characterisation is the integration of the three previous steps (hazard identification, exposure assessment, hazard characterisation) to obtain a *risk estimate* of the likelihood and the severity of the adverse effects in a given population with attendant uncertainties.

Risk characterisation is the final stage of risk assessment. This can be qualitative (low, medium, high) or quantitative (number of human

Table 3.10   Predicted frequency of *Salmonella* infection following ingestion of 100 *Salmonella* cells by two 100 000 populations using a mechanistic model (summarised from Buchanan *et al.* (2000)).

	Immune status	Adults <65 years old	Adult >65 years old
Number of individuals infected	Immunocompetent	3297	13910
	Immunoimpaired	33	2669
	Total	3330	16579
Number of symptomatic individuals	Immuncompetent	165	696
	Immunoimpaired	3	267
	Total	168	963
Number of deaths	Immunocompetent	3.3	13.9
	Immunoimpaired	0.2	21.3
	Total	3.5	35.2

infections, illnesses or deaths per year or per 100 000 population), depending upon the exposure assessment. The degree of confidence in the risk estimate depends on the amount of knowledge in the previous steps; for example, the variation in the human (sub)population suscept-ibility. While quantitative estimates of risk are highly desirable, they are difficult to obtain owing to limitations in expertise, time, data and meth-odology. In risk characterisation the variation is divided into '*uncertainty*' which reflects where important data is not available and '*variability*' where the data is not constant due to recognised factors such as variable amount of food eaten and population susceptibility. These must be dis-tinguished in modelling (Nauta 2000).

The overall probability, or risk estimate, is determined from

Risk estimate = Dose response assessment × Exposure assessment

This is shown schematically in Fig. 3.10. Where the exposure assessment is the input value(s) for the dose–response relationship, the result is the risk estimate, i.e. the probability of an adverse effect.

The risk characterisation should not only determine the relative risk that a hazard will pose to a population, but also its severity. Hence the use of multiple end-points, such as infection, illness (morbidity) and death (mortality).

Unfortunately, scientific data are frequently unavailable for parts of the risk assessment, so any uncertainties and variations should be noted to assist in making the process transparent. Additionally, any changes in the process would require a reassessment of the risk. Hence accurate risk

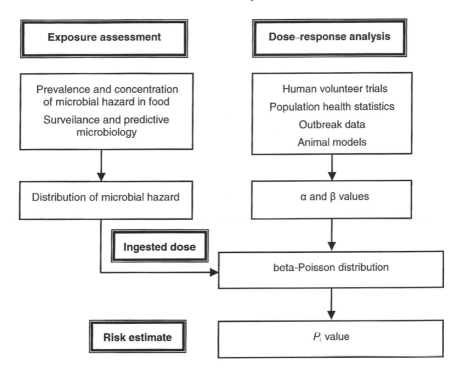

Fig. 3.10 Predicting the probability of infection.

communication (including uncertainties) is a very important aspect of risk analysis (Section 3.6).

Risk characterisation estimates can be assessed by comparison with independent epidemiological data that relate hazards to disease prevalence. Using probabilistic models, Monte Carlo simulations can be used to modify previous assumptions and values to ascertain their relative importance. Important variables are frequently illustrated through tornado chart presentation (Section 3.3.10). A risk management strategy can then be formulated from the risk characterisation.

3.3.9 Production of a formal report

The risk assessment should be fully and systematically documented. To ensure transparency, the final report should indicate, in particular, any constraints and assumptions relative to the risk assessment. The report should be made available to independent parties on request.

3.3.10   *Triangular distributions and Monte Carlo simulation*

Risk assessments need to separate 'uncertainty' (due to lack of knowledge) from 'variability' (due to known factors such as biological variation), and must be described in a transparent fashion. There are two types of quantitative risk assessments, 'deterministic' and 'stochastic'. Deterministic models use single point estimates as input data, whereas stochastic models use a distribution range of data values. To determine the risk estimate, the input data may be point estimates of the average value or worst case (95th percentile). In contrast to point estimates, probability distributions describe the relative weightings of each possible value and are characterised by a number of parameters: minimum, mostly likely and maximum. By describing these three values, a triangular distribution is generated which can be further analysed using Monte Carlo simulation (see below). An example of triangular distribution ($-1.0, 0.5, 2.5$) is given in Fig. 3.11 for the effect of evisceration of poultry on microbial load. There is no log reduction in microbial numbers, and the most likely effect was a $0.5\log_{10}$ increase. The problem with constructing a triangular distribution (as shown in Fig. 3.11) is that it does not fully reflect the 'crude data' distribution. Nevertheless, the probability distribution produces a risk distribution that characterises the range of risk of a population. The stochastic method is also known as the 'probabilistic approach'.

The Monte Carlo process is a procedure that generates values of a random variable based on one or more probability distributions (Vose 1996). Monte Carlo simulation is a model that uses the Monte Carlo pro-

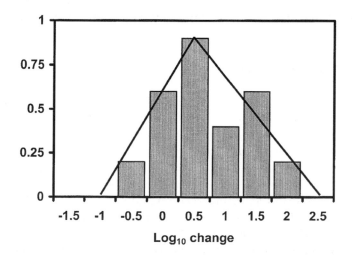

Fig. 3.11   Triangular distribution of *C. jejuni* on poultry carcasses after evisceration (Fazil *et al.* 2000b).

cess to calculate a model output value many times with different input values. The purpose is to obtain a complete range of all possible scenarios. In microbiological risk assessment there may be two or more variables, such as prevalence of the pathogen and the concentration of pathogen, which are multiplied together and hence generate a further probability range for further calculations (Vose 1997, 1998).

Figure 3.12 shows the frequency distribution for three variables. There are three ways of assessing the data:

(1) Determine the mean of each distribution and multiply them together
(2) Take the highest value for each dataset to determine the worst case scenario
(3) Use Monte Carlo simulation to take random values from each dataset after repeated sampling (several thousand times) and determine a distribution curve of likely results.

The third approach is obviously more representative of the situation.

A convenient means of Monte Carlo simulation is to first enter the variable ranges in Excel™ (Microsoft Corp.) spreadsheet files and then use either @RISK (Palisaide Corp.), Crystal Ball 2000 (Decisioneering) or Analytica (Lumina Decision Systems, Inc.) which are risk analysis add-in tools (see Internet Directory for contact details) to calculate the resultant distribution.

An example of illustrating risk mitigation factors using a tornado chart (so called due to its shape) is given in Fig. 3.13 for *V. parahaemolyticus* (FDA 2000b; Section 4.7.1). The purpose is to easily communicate the major factors which might be altered in risk-mitigating strategies.

3.4 Risk management

Risk management is required when epidemiological and surveillance data demonstrate that specific foods are possible hazards to consumer health due to the presence of pathogenic micro-organisms or microbial toxins. Governmental risk managers must decide on appropriate control options to manage this risk. To understand the risk for consumers more explicitly, the risk managers may initiate a microbiological risk assessment. This assessment leads to an estimate of human risk together with associated uncertainty and variability limits. Risk managers must be aware of these limitations when considering risk management options. A separation between risk management and risk assessment activities must be maintained for the assessment to be transparent.

Following risk assessment, appropriate risk management steps should

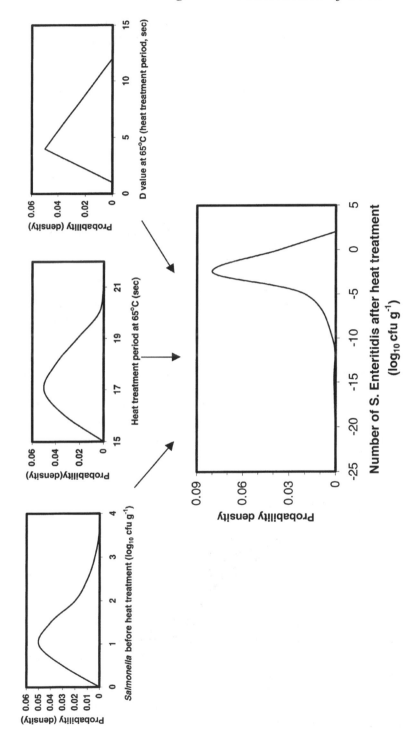

Fig. 3.12   Monte Carlo simulation of *S. Enteritidis* distribution after heat treatment.

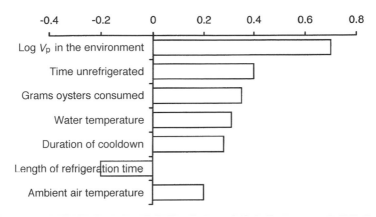

Fig. 3.13 Risk factors for *V. parahaemolyticus* in raw molluscan shellfish (FDA 2000b).

result in safe handling procedures and practices, food processing quality and safety assurance controls, and food quality and safety standards and criteria (Notemans *et al.* 1995). If required, risk managers may select and implement appropriate regulatory measures. A guidance document on the interaction between assessors and managers of microbiological hazards in foods has been developed (WHO/FAO 2000b). The international risk manager for food is the CAC. A diagram of risk management activities is given in Fig. 3.14.

Risk management can be divided into four aspects:

(1) Risk evaluation: initial risk management activities
(2) Risk management option assessment
(3) Implementation and management of decisions
(4) Monitoring and review.

The general principles are as follows:

(1) Risk management should follow a structured approach.
(2) Protection of human health should be the primary consideration in risk management decisions.
(3) Risk management decisions and practices should be transparent.
(4) Determination of risk assessment policy should be a specific component of risk management.
(5) Risk management should ensure the scientific integrity of the risk assessment process by maintaining a functional separation of risk management and risk assessment.

Fig. 3.14   Risk management activities.

(6)   Risk management decisions should take into account the uncertainty in the output of the risk assessment.
(7)   Risk management should include clear, interactive communication with consumers and other interested parties in all aspects of the process.
(8)   Risk management should be a continuing process that takes into account all newly generated data in the evaluation and review of risk management decisions.

Risk limitation has two principles:

(1)   The individual risk resulting from a risk source should not exceed the maximum permissible level.
(2)   The risk is to be reduced 'as low as reasonably achievable' and correct implementation of optimisation must be demonstrated.

A specific aspect of this approach relates to the criteria utilised and the related values. With regard to the maximum permissible level and to the negligible level, where this approach has been applied, the criterion utilised is frequently the lifelong risk of death. The figures range from 10^{-8} to 4×10^{-3}, though most figures converge between 10^{-6} and 10^{-4}. For the food industry, the maximum permissible level of risk may also be translated into an expression of the maximum level and/or frequency of a hazard, e.g. in a given product and termed 'food safety objective' (Section 3.5).

The SPS Agreement (Section 1.7.1) refers to the 'appropriate level of sanitary or phytosanitary protection' (ALOP). This is defined as the level of protection deemed appropriate by the WTO member establishing a sanitary or phytosanitary measure to protect human, animal or plant life or health within its territory. This assumes a threshold value dividing an unacceptable risk from an acceptable risk. The CCFH (2000) proposes a 'tolerable risk', again with a dividing threshold value. The term 'food safety objective' (Section 3.5) is used in some CAC documents, though it has not yet been clearly defined. The ICMSF defines a food safety objective as a statement of the frequency or maximum concentration of a microbiological hazard in a food considered acceptable for consumer protection (ICMSF 1996b, 1998b,c). (See also Section 2.8 on microbiological criteria.) 'As low as reasonably achievable' (ALARA) is a concept for risk management which does not have an absolute value dividing acceptable and unacceptable (tolerable) risk. Instead risk is categorised into three bands: intolerable, tolerable and acceptable (Fig. 3.15). Intolerable (unacceptable) risks are managed by regulations and bans. The goal of risk management is to use strategies to reduce the risk to ALARA and may be a

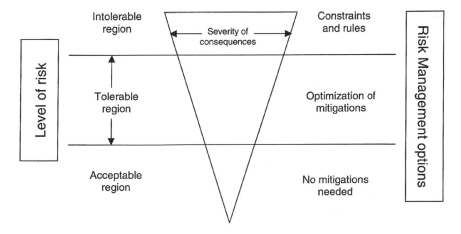

Fig. 3.15 As low as reasonably achievable (ALARA) approach to risk management (Jouve 2001).

more appropriate approach than defining a threshold value between acceptable and unacceptable.

HACCP (Section 2.3) is a risk management system previously developed on the basis of qualitative risk assessment. Two key advantages that now come from using a quantitative approach are the ability to link the HACCP plan to an estimate of public health impact and a measure of the level of confidence that the evaluators have in their results.

Risk assessment for microbiological hazards is partially derived from chemical risk assessment. However, microbial contaminants are a significantly more complex group of hazards than chemicals. The analysis must include the estimation of microbial growth rates under plausible conditions of food processing and storage. Because this is difficult due to lack of data, the main approach has been to identify places in the processing chain that either lead to microbial contamination or allow microbial growth. At its present state of development, the HACCP approach of risk management does not use microbial dose–response information. It uses performance standards (numbers of microbes per gram of food) which are practical rather than scientific standards. Eventually, however, these standards will have to be justified because world trade regulations will require it.

3.4.1   Risk assessment policy

Guidelines for value judgement and policy choices which may need to be applied at specific decision points in the risk assessment process are known as the risk assessment policy. Risk assessment policy setting is a risk management responsibility, which should be carried out in full collaboration with risk assessors, and which serves to protect the scientific integrity of the risk assessment. The guidelines should be documented to ensure consistency and transparency. Examples of risk assessment policy setting are, establishing the population(s) at risk, establishing criteria for ranking hazards and guidelines for application of safety factors.

3.4.2   Risk profiling

Risk profiling is the process of describing a food safety problem and its context, to identify those elements of the hazard or risk relevant to various risk management decisions. The risk profile would include identifying aspects of hazards relevant to prioritising and setting the risk assessment policy and aspects of the risk relevant to the choice of safety standards and management options. A typical risk profile might include the following:

- A brief description of the situation, product or commodity involved
- Identification of what is at risk, e.g. human health, economic concerns, potential consequences, consumer perception of the risks
- Description of risks and benefits.

A semi-quantitative risk profile approach has been proposed (CCFRA 2000) which offers considerable help in microbiological risk assessment (see also Voysey & Brown 2000). The Danish Veterinary and Food Administration constructed a risk profile of *C. jejuni* (Section 4.3.2) which was further utilised in the construction of a risk assessment of human illness to *C. jejuni* in broilers (Section 4.3.3).

3.5 Food safety objectives

The next step after risk assessment and food safety management involves establishing a food safety objective, i.e. a statement of the maximum level of a microbiological hazard in a food considered acceptable for consumer protection. The food safety objective is a risk management tool linking risk assessment and effective measures to control identified risks. For practical implementation in specific sectors of the food chain, it is the responsibility of governmental authorities to translate the output of risk analysis into food safety objectives. Such objectives delineate the specific target(s) that any food operator concerned should endeavour to achieve through appropriate interventions.

Food safety objectives are a statement of the maximum level of a microbiological hazard in a food considered acceptable for human consumption. An objective should

- Be technically feasible
- Include quantitative values
- Be verifiable
- Be developed by governmental bodies with a view to reaching consensus for a food in international trade.

Food safety objectives as defined by governmental authorities represent the minimum target on which food operators base their own approach (see Fig. 3.1). The government's food safety objectives may be adopted as a company's food safety requirements. Alternatively, depending on commercial factors, a company may wish to establish more demanding food safety requirements which input into their food safety programme. Achieving the food safety requirements will require the implementation of Good Manufacturing Practice (GMP), Good Hygienic Practice, HACCP and Quality Assurance systems (Section 2.3).

Microbiological criteria such as food safety objectives could be based upon established methods of certification, inspection, and/or microbiological testing. The establishment of microbiological criteria should consider:

- Evidence of actual or potential hazards to health
- Microbiology of the raw materials
- Effects of processing
- Likelihood and consequences of contamination and growth during handling, storage and use
- Category of consumers at risk
- The distribution system and potential for consumer abuse
- The reliability of the methods used to determine product safety
- Cost/benefit ratio of the application
- Intended use of the food.

Microbiological criteria are used to ensure the safety of food, adherence to GMPs, the keeping quality of certain perishable foods and/or the suitability of a food or ingredient for a particular purpose. Microbiological criteria, when appropriately applied, can be a useful means for ensuring safety and quality of foods, which in turn, elevates consumer confidence. It also can provide the food industry and regulatory agencies with guidelines for control of food processing systems. Internationally accepted criteria can advance free trade through standardisation of food safety and quality requirements. However, it should be recognised (as shown in Section 2.8) that sampling plans have inherent risks associated with them (see Figs 2.8 and 2.9).

3.6   Risk communication

Risk communication is an exchange of information and opinions throughout the risk analysis process, concerning risk, risk-related factors and risk perception, among all interested parties, to explain risk assessment findings and the basis of risk management decisions.

Goals of risk communication
- Promote awareness and understanding by all participants of the specific issues under consideration during the risk analysis process
- Promote consistency and transparency in arriving at and implementing risk management decisions
- Provide a sound basis for understanding the risk management decisions proposed or implemented

- Improve the overall effectiveness and efficiency of the risk analysis process
- Contribute to the development and delivery of effective information and education programmes when they are selected as risk management options
- Strengthen the working relations and mutual respect among participants
- Promote the appropriate involvement of all interested parties in the risk communication process
- Exchange information on the knowledge, attitudes, values, practices and perceptions of interested parties concerning risk associated with food and related topics
- Foster public trust and confidence in the safety of the food supply chain.

As a result of numerous food scares and well publicised food poisoning outbreaks, the public have become increasingly concerned about the risks associated with food (Table 1.7). Hence effective risk communication with consumers is both important and necessary. Risk communication is required to adequately address and respond to needs for criteria, hazards, risks, safety, and general concerns about food. Risk communication provides the public with results of expert scientific review of food hazard identification and assessment of risk to the general population or target group. It also provides the private and public sectors with information necessary to prevent, reduce and minimize food risks through systems of quality and safety. Additionally, it is essential that risk communication provides sufficient information for populations at greatest risk in terms of any particular hazard to exercise their own options to achieve protection.

Aspects of risk communication

(1) Nature of the risk
- Characteristics and importance of hazard of concern
- Magnitude and severity of risk
- Urgency of situation and trend
- Probability of exposure
- Amount of exposure that constitutes a significant risk
- Population at risk.

(2) Nature of the benefits
- Benefits associated with risk
- Who benefits and in what way?
- Balance point between risk and benefit
- Magnitude and importance of benefit
- Total benefit to all affected populations.

(3)   Uncertainties in risk assessment
- Methods used to assess risk
- Importance of each of uncertainties
- (In)accuracy of available data
- Assumptions on which the estimates are based
- Effect of changes in the estimates on risk management decisions.

(4)   Risk management options
- Action(s) taken to control/manage the risk
- Actions individuals may take to reduce individual risk
- Justification for choosing a specific risk management option
- Benefit(s) of a specific option
- Who benefits?
- Cost of managing a risk – who pays?
- Risks that remain after a risk management option is implemented.

The differences in the public's perception of 'risk' and 'benefit' need to be understood, and also how these perceptions vary between countries and social groups. For example, there are marked differences in the acceptance of genetically modified food ingredients between the UK and USA, partly due to public distrust of reassurances from politicians and food experts since the emergence of BSE-vCJD (Bruce *et al.* 1997; Gale *et al.* 1998; Ferguson *et al.* 1999). There are differences in risk perception between men and women. In general, women tend to perceive more risk from technological and food-related hazards than do men. Social inclusion is also likely to improve trust in government and, by implication, the regulatory framework associated with risks. Increasing transparency in risk management processes, and the need to improve public trust in those processes, are likely to involve increased public participation in risk management itself.

4

APPLICATION OF
MICROBIOLOGICAL RISK
ASSESSMENT

4.1   Introduction

There are a growing number of published microbiological risk assessments, some of which have been focused on one aspect, i.e. hazard characterisation (see Tables 1.10 and 3.2). Those by the JEMRA (Section 3.2) contain an extensive literature review of the subject matter, including unpublished data. Therefore only essential aspects of the risk assessments are given here. The microbiological risk assessments published to date vary considerably in the depth of assessment and also structure. The reader should appreciate that these examples do not always use the *Codex Alimentarius* structure and definitions for risk assessment (see Section 1.8.2, Fig. 1.4), and that this is an evolving scientific discipline. The following sections are not an exhaustive survey of published microbiological risk assessments but are abridged examples to give an indication of the processes involved. Each section starts with a resumé of the target organism as a form of hazard identification.

4.2   *Salmonella* spp.

Sources:

- Domestic and wild animals: poultry, pigs, cattle, rodents, cats and dogs
- Infected humans (especially *S.* Typhi and *S.* Paratyphi).

Control:

- Heat treatment (pasteurisation, sterilisation)
- Refrigeration
- Prevention of cross-contamination
- Good personal hygiene
- Effective sewage and water treatment processes.

Salmonella is a genus of the *Enterobacteriaceae* family and hence comprises Gram-negative, facultative anaerobic bacteria. Their optimum growth temperature is about 38°C and their minimum growth temperature is about 5°C (Table 2.4). Because they do not form spores they are relatively heat sensitive, being killed at 60°C in 15–20 minutes ($D_{62.8}$ = 0.06 min). Of current concern is the increase in multiple-antibiotic-resistant serotypes. Plasmids in the range 50–100 kb have been associated with *Salmonella* virulence (Slauch *et al.* 1997).

There are only two species of *Salmonella* (*S. enterica* and *S. bongori*) which are divided into eight groups (Boyd *et al.* 1996). This classification is useless with regard to epidemiological investigations and therefore detailed characterisation is required. The genus *Salmonella* contains over 2324 different 'strains' which are also called serovars or serotypes. These are distinguished by their O-, H- and Vi-antigens using the Kaufmann–White scheme (Ewing 1986). The serotypes are then put into serogroups according to common antigenic factors. Members of the genus *Salmonella* have a complex lipopolysaccharide (LPS) structure (Mansfield & Forsythe 2001) which gives rise to the O-antigen. It is the serotype of the *Salmonella* isolate that aids epidemiological studies to trace the vector of *Salmonella* infections.

Characteristic symptoms of *Salmonella* food poisoning are as follows:

- Watery diarrhoea
- Nausea
- Abdominal pain
- Mild fever and chills
- Sometimes vomiting, headache and malaise.

Thorns (2000) estimated the incidence of salmonellosis (per 100 000) as 14 (USA), 38 (Australia), 73 (Japan), 16 (Netherlands) and 120 in parts of Germany. The incubation period before illness is generally between 8 and 72 hours. The illness is usually self-limiting, lasting 4–7 days, and most people fully recover without medical treatment. Occasionally systemic infections will occur, often due to *S.* Dublin and *S.* Choleraesuis, which may require fluid and electrolyte replacement treatment. The infected person will be shedding large numbers of salmonellae in their faeces during the period of illness (average of 5 weeks). The numbers of salmonellae in the faeces will then decrease but they may persist for up to 3 months; approximately 1% of cases become chronic carriers. Children excrete up to 10^6-10^7 salmonellae per gram of faeces during convalescence.

Chronic consequences include postenteritis reactive arthritis, and Reiter's syndrome may follow 3–4 weeks after the onset of acute symptoms. Reactive arthritis may occur in 1–2% of cases. Reactive arthritis and Reiter's syndrome are rheumatoid diseases caused by a range of bacteria which induce septic arthritis by haematogenous spread to the synovial space, causing inflammation. Causative organisms include *S.* Enteritidis, *S.* Typhimurium and other serotypes such as *S.* Agona, *S.* Montevideo and *S.* Saint Paul. Non-salmonella bacteria causing reactive arthritis include *C. jejuni* (see Section 4.3), *Sh. flexneri*, *Sh. sonnei*, *Y. enterocolitica* (in particular O:3 and O:9), *Y. pseudotuberculosis*, *E. coli* and *K. pneumoniae*. These conditions are related to a genetically determined host risk factor, the major histocompatibility complex (MHC) gene for the Class 1 antigen, HLA-B27 and cross reaction with bacterial antigen leading to an autoimmune anti-B27 response. Those who are human leukocyte antigen HLA-B27 positive have an 18-fold greater risk for reactive arthritis, a 37-fold greater risk for Reiter's syndrome, and up to a 126-fold greater risk for ankylosing spondylitis than persons who are HLA-B27 negative and have the same enteric infections. However, a lack of correlation between reactive arthritis and HLA-B27 has been reported after *S.* Typhimurium and *S.* Heidelberg/*S.* Hadar outbreaks in Canada (Thomson *et al.* 1995). It is probable that other genes which may be related or act in concert determine which disease is acquired. The condition is immunological; hence patients do not benefit from treatment with antibiotics, but are treated with non-steroidal anti-inflammatory drugs.

Infective dose varies according to the age and health of the victim, the food and also the *Salmonella* strain. The infectious dose (100% probability of infection) varies from 20 cells to 10^6 cells according to serotype (see Section 4.2.2 for further discussion), food and vulnerability of host. It should be noted that the first 50 ml of liquid passes straight through the stomach into the small intestines and is therefore protected from the stomach's hostile acidic environment. Likewise, it is believed that chocolate can protect *Salmonella* while transient in the stomach, hence reducing the infectious dose. The disease is caused by the penetration and passage of *Salmonella* organisms from the gut lumen into the epithelium of the small intestine where they multiply. Subsequently, the bacteria invade the ileum and even occasionally the colon. The infection elicits an inflammatory response. The number of salmonellosis cases shows a marked seasonal trend, with peak incidences in the summer.

Salmonella infections in animals generally differ from the typical gastroenteritis and other sequelae produced in humans (Berends *et al.* 1997). Therefore the use of animal models is limited. However, unlike most other bacterial pathogens, there is a considerable amount of human data on salmonellosis.

Ninety-six per cent of cases are estimated to be caused by a wide range of contaminated foods (see Table 1.2). This includes raw meats, poultry, eggs, milk and dairy products, fish, shrimp, frog legs, yeast, coconut, sauces and salad dressing, cake mixes, cream-filled desserts and toppings, dried gelatin, peanut butter, cocoa and chocolate. Contamination of the foods is through poor temperature control and handling practices, or cross-contamination of processed foods from raw ingredients. The organism multiplies on the food to an infectious dose.

In addition to contaminating egg shells, *S.* Enteritidis can be isolated from the egg yolk due to transovarian infection. The organism travels up the anus from the environment and colonises the ovaries. *S.* Enteritidis subsequently infects the egg before the protective shell is formed. An infected unfertilised egg will result in contaminated egg products, whereas a fertilised egg results in a chronically ill chick with systemic infection and hence a contaminated carcass.

S. typhi and *S. paratyphi* A, B, and C produce typhoid and typhoid-like fever in humans. Typhoid fever is a life-threatening illness. The organism multiplies in the submucosal tissue of the ileal epithelium and then spreads throughout the body via macrophages. Subsequently, various internal organs such as the spleen and liver become infected. The bacteria infect the gall bladder from the liver and finally infect the intestines using bile as the transportation medium. If the organism does not progress past the gall bladder, then no typhoid fever develops. Nevertheless, the person may continue to shed the organism in their faeces. Typical symptoms of typhoid fever are as follows:

• Sustained fever as high as 39–40°C
• Lethargy
• Abdominal cramps
• Headache
• Loss of appetite
• Rash of flat, rose-coloured spots may appear.

The fatality rate of typhoid fever is 10% compared to less than 1% for most forms of salmonellosis. A small number of people recover from typhoid fever but continue to shed bacteria in their faeces. *S.* Typhi and *S.*

Paratyphi enter the body through food and drinks that may have been contaminated by a person who is shedding the organism in their faeces. Seventy per cent of cases are associated with foreign travel.

4.2.1   S. Enteritidis in shell eggs and egg products

The Food Safety and Inspection Service (FSIS 1998) completed a 2 year comprehensive risk assessment of *S.* Enteritidis in shell eggs; this can be accessed from the Web (see Table 1.10 and Internet Directory for the URL; the document is 268 pages long). The *S.* Enteritidis risk assessment (SERA) model can also be downloaded. It requires Excel™ (Microsoft Corp.) and @RISK (Palisade Corp.) software to run. The model is in many ways the archetypal risk assessment for food-borne pathogens and is frequently referred to by the other *Salmonella* risk assessments. There is an online study guide to the SERA model by Wachsmuth (see Internet Directory). The consequence of risk models such as FSIS (1998) is the action plan in the USA to eliminate *S.* Enteritidis illness (Anon. 1999c).

The objectives of the risk assessment were to:

- Identify and evaluate potential risk reduction strategies
- Identify data needs
- Prioritise future data collection efforts.

The risk assessment model consists of five modules (Fig. 4.1).

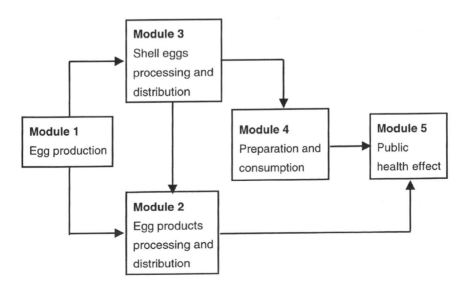

Fig. 4.1 'Farm-to-fork' risk assessment model for shell eggs and egg products (FSIS 1998).

Egg production module

This estimates the number of eggs produced that are infected (or internally contaminated) with S. Enteritidis.

Shell egg processing and distribution module

This module follows the shell eggs from collection on the farm through processing, transportation and storage. The eggs remain intact throughout this module. Therefore the primary factors affecting the level of S. Enteritidis are the cumulative temperatures and times of the various processing, transportation and storage stages. The two important modelling components are the time until the yolk membrane loses its integrity (and hence barrier to S. Enteritidis) and the subsequent growth rate of S. Enteritidis in eggs.

The lag period before yolk membrane breakdown time (YMT) was estimated from the equation

$$\log_{10} YM = \{2.08 - 0.04257 * T) \pm (2.042 * 0.15245)[(1/32)$$
$$+((T - 21.6)^2/(32 * 43.2))]^{0.5}\}$$

The subsequent growth rate was estimated from

$$\text{Growth rate } (\log_{10} \text{cfu h}^{-1}) = -0.1434 + 0.026 * \text{internal}$$
$$\text{egg temperature } (^\circ C)$$

Egg products processing and distribution module

This module tracks the change in numbers of S. Enteritidis in egg processing plants from receipt of eggs through pasteurisation. The death rate of S. Enteritidis in whole eggs and yolk during pasteurisation is determined from experimentally derived D values (Fig. 2.5). There are two sources of S. Enteritidis in egg products: the internal contents of eggs and cross-contamination during breaking.

Preparation and consumption module

This estimates the increase or decrease in the numbers of S. Enteritidis organisms in eggs or egg products as they pass through storage, transportation, processing and preparation.

The public health module

This calculates the incidences of illnesses and four clinical outcomes (recovery without treatment, recovery after treatment by a physician, hospitalisation and mortality) as well as the cases of reactive arthritis associated with consuming S. Enteritidis-positive eggs (Table 4.1).

Table 4.1   Public health module results for *S.* Enteritidis risk assessment (FSIS 1998).

	Category	Mean
Normal population	Exposed	1 889 200
	III	448 803
	Recover with no treatment	425 389
	Physician visit and recovery	21 717
	Hospitalised and recovered	1 574
	Death	123
	Reactive arthritis	13 578
Susceptible population	Exposed	521 705
	III	212 830
	Recover with no treatment	196 295
	Physician visit and recovery	14 491
	Hospitalised and recovered	1 776
	Death	269
	Reactive arthritis	6 416
Total population	Exposed	2 410 905
	III	661 633
	Recover with no treatment	621 684
	Physician visit and recovery	36 208
	Hospitalised and recovered	3 350
	Death	392
	Reactive arthritis	19 994

The results of the baseline model predict:

- Average production of 46.8 billion shell eggs per year (in the USA)
- 2.3 million eggs containing *S.* Enteritidis
- This results in 661 633 human illnesses per year; of those taken ill
  94% recover without medical care
  5% visit a physician
  0.5% are hospitalised
  0.05% die.
- 20% of the population are considered to be at a higher risk: infants, elderly, transplant patients, pregnant women, individuals with certain diseases.

The output of the module was validated using egg culturing data from California to predict that there were 2.2 million *S.* Enteritidis contaminated eggs in the USA each year. The model was also validated using public health surveillance data.

The beta-Poisson model (Section 3.3.6) from *Salmonella* human volunteer feeding trials (1930–1973) estimates a probability of infection of 0.2 from ingesting 10^4 *Salmonella* cells (Fig. 3.8). Because an infectious dose does not necessarily lead to illness, the probability of infection is greater than the probability of illness. These data were obtained using serotypes other than *S.* Enteritidis and hence were accepted as not being totally appropriate. They are further discussed by Fazil *et al.* (2000a) and are summarised in Section 4.2.2.

The baseline egg products model predicts that the probability is low that any cases of *S.* Enteritidis infection will result from the consumption of pasteurised egg products. However, the current FSIS time and temperature regulations do not provide sufficient guidance to the egg products industry for the large range of products the industry produces (Table 4.2). Time and temperature standards based on the amount of bacteria in the raw product, how the raw product will be processed, and the intended use of the final product will provide greater protection to the consumers of egg products.

Table 4.2 USDA minimum time and temperature requirements for three egg products.

Liquid egg product	Minimum temperature requirements		Minimum holding time requirements (minutes)
	°F	°C	
Albumen	134	56.7	3.5
	132	55.5	6.2
Whole egg	140	60	3.5
Plain yolk	142	61.1	3.5
	140	60	6.2

The percentage reduction for total human illnesses was calculated for two scenarios differing from current practice within the Shell Egg Processing and Distribution module. The first scenario was that if all eggs were immediately cooled after laying to an internal temperature of 7.2°C (45°F) and then maintained at this temperature, a 12% reduction in human illnesses would be the result. Similarly an 8% reduction in human illnesses would be the result if eggs were maintained at an ambient (i.e. air) temperature of 7.2°C (45°F) throughout shell egg processing and distribution.

4.2.2 Hazard identification and hazard characterisation of Salmonella *in broilers and eggs*

The draft JEMRA report entitled 'Hazard identification and hazard characterisation of *Salmonella* in broilers and eggs' of Fazil *et al.* (2000a) contains an extensive amount of information (in common with other JEMRA reports) which can be downloaded from the Internet (see Internet Directory for URL; the document is 110 pages in length). The first section describes the public health outcomes, pathogen characteristics, host characteristics and food-related factors that may affect the survival of *Salmonella* in the human intestinal tract (Table 2.4). The second section reviews three dose–response models for salmonellosis and compares the results with 33 sets of outbreak data (Section 3.3.6). Where possible, differences in the dose–response are characterised for susceptible and normal subgroups of the population.

The three dose–response models reviewed by Fazil (2000a) are:

(1) *S.* Enteritidis in eggs from FSIS (1998). This is a beta-Poisson model derived using the results of feeding studies of *Sh. dysenteriae*, with illness as the biological end-point.
(2) Health Canada model (unpublished, quoted by Ebel *et al.* (2000)) based on the Weibull function (Section 3.3) which was derived from human feeding studies for several different bacterial pathogens and data from two *Salmonella* outbreaks (Paoli, unpublished report; Ross, unpublished report).
(3) Beta-Poisson model derived from human feeding study data of prisoners (McCullough & Eisele 1951a–c).

The feeding trial data from McCullough & Eisele 1951a,b which used *S.* Anatum, *S.* Bareilly, *S.* Derby, *S.* Meleagridis and *S.* Newport are shown in Fig. 4.2. The data has been corrected (as per Fazil *et al.* 2000a) for subjects who had received multiple doses and hence may have acquired some immunity. The appropriate beta-Poisson fit is also shown.

The epidemiological models of FSIS (1998) and Health Canada (unpublished) were subdivided by Fazil *et al.* (2000a) into normal and susceptible subgroups and are shown in Fig. 4.3. The beta-Poisson fit from Fig. 4.2 is also included for comparison.

Data from 33 outbreaks of salmonellosis was compiled by Fazil *et al.* (2000a) and used for comparison with the previous dose-response curves. Use was made of data collected in Japan where large-scale facilities (> 750 meals per day or > 300 dishes of a single menu) have been advised to save 50 g portions for a minimum of 2 weeks at −20°C for future examination in the case of illness being associated with the food. Because

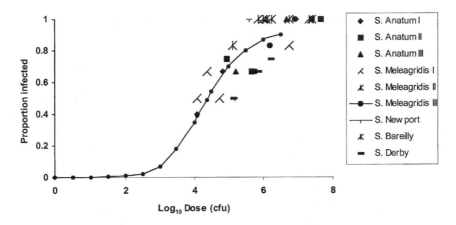

Fig. 4.2   Human volunteer feeding trials for *Salmonella* serovars (from Fazil *et al.* (2000a) using the data of McCullough & Eisele (1951a,b)).

S. Enteritidis is the major cause of *Salmonella* food poisoning, the majority of outbreaks were from this serovar; this contrasts with the feeding data where a number of serovars were tested. The outbreak data are shown in Fig. 4.4. The data was fitted to the three models previously described. None of the models best fitted the outbreak data over the full dose range;

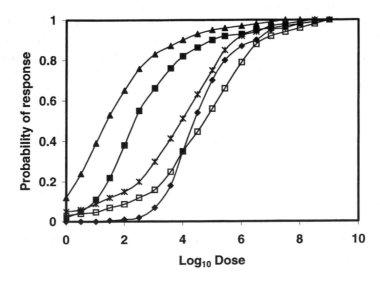

Fig. 4.3   Epidemiological models of FSIS (1998) and Health Canada (Fazil *et al.* 2000a). Health Canada: □, normal; ✕, susceptible. FSIS: ■, normal; ▲, susceptible. ◆, Naïve beta-Poisson (Fig. 4.2).

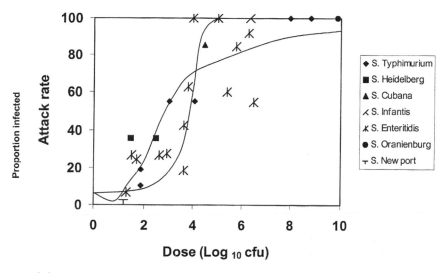

Fig. 4.4   Outbreak dose–response curve (Fazil *et al.* 2000a).

the naïve beta-Poisson model, in particular, significantly underestimated the probability of illness.

Overall it was estimated that children under 5 years of age were 1.8–2.3 times more likely to become ill with the same ingested dose as the normal population. There was no evidence to indicate that *S.* Enteritidis was more infectious than another serovar at the same ingested dose. Hence a common dose–response model for all non-typhoid and non-paratyphoid *Salmonella* spp. can be used. The report did not consider secondary transmission, chronic sequelae such as reactive arthritis, or the effect of the food matrix.

4.2.3 *Exposure assessment of Salmonella Enteritidis in eggs*

Ebel *et al.* (2000) have released an exposure assessment of *S.* Enteritidis in eggs. Although this can be downloaded from the Web (see Internet Directory) the manuscript carries the statement 'Draft: Do Not Cite or Quote' on every one of its 119 pages. Hence no data are reproduced here. Nevertheless, the document compares three previous exposure assessments:

(1)   FSIS (1998)
(2)   Whiting & Buchanan (1997)
(3)   Health Canada (unpublished).

The first two are available to the reader and can be used to follow the changes in *S.* Enteritidis prevalence and numbers through a 'farm-to-fork' process as outlined in Fig. 3.3.

4.2.4   Exposure assessment of Salmonella spp. in broilers

Kelly *et al.* (2000) have released the fullest exposure assessment to date of *Salmonella* spp. in broilers. This includes products that are fresh, frozen and further processed. The preliminary report of length 104 pages is available for downloading (see Internet Directory) and contains an extensive literature review. The report offers a model framework for future risk assessments and highlights current limitations in the information available. As previously stated, although *Salmonella* is a well studied food poisoning bacterium, current approved microbiological criteria require enumeration on a presence or absence basis (less than one *Salmonella* cell 25 g^{-1}); hence there is a lack of quantitative data on numbers of *Salmonella* in the primary source (i.e. poultry) and in food. There is an overlap in approach with the risk assessment for *C. jejuni* in fresh chicken (Section 4.3.1). However, *C. jejuni* is more temperature-sensitive than *Salmonella* and does not grow at temperatures below 30°C, but does have a reportedly higher infectivity (Table 2.4, Section 4.3).

The framework is similar to that shown in Fig. 3.3, although the fourth module has a slightly different name.

Production module

This module estimates the prevalence of *Salmonella*-positive broilers leaving the farm. Data required include the source of infection, prevalence of *Salmonella* in the flock, the number of *Salmonella* per positive bird and sampling methodology (see Section 3.3.7 re. dose estimate). Various epidemiological and farm management factors will influence these values, and currently there is little quantitative data on the number of *Salmonella* per bird.

Transport and processing module

This module ultimately aims to estimate the prevalence and concentration of *Salmonella* at the end of processing. Therefore it needs estimations of prevalence and concentration at each step of processing (cf. *C. jejuni*, Section 4.3.1). An estimation of cross-contamination during transportation and processing needs to be included, although currently there is little data on this topic (see Christensen *et al.* 2001, Section 4.3.3).

Retail, distribution and storage module

This module estimates the changes in prevalence and concentration between processing and consumer preparation. Periods when the temperature supports microbial multiplication on contaminated meat need to be determined to estimate the growth and persistence of the pathogen. Temperature abuse can occur in both the retail sector and the consumer aspect of this module. Future studies collecting relevant data for predictive microbiology models are required to improve this module (see Section 2.7).

Preparation module
The changes in numbers of *Salmonella* due to preparation, including cross-contamination, are considered in this module. Improved predictive models are required with regard to the thawing (thermal profiles) of frozen contaminated carcasses as well as death rates (*D* values) due to cooking (Section 2.5.4, Fig. 2.5). The outputs from this module are an estimate of the prevalence of contaminated products and the number of ingested *Salmonella* cells.

Consumption pattern module
This module requires data about the consumers. This includes not only age, sex and immunological status, but also behaviour which can be age and nationality related. Most information available is based on 'average consumption per day' and does not describe portion size or frequency of consumption. The amount ingested by different subpopulations (i.e. normal and susceptible) is estimated and combined with the output from the Preparation module to generate an overall estimate of exposure.

The estimate of exposure can be used through dose–response analysis (hazard characterisation) to determine the probability of both infection and illness (Section 3.3.6).

4.2.5   Salmonella *spp. in cooked chicken*

Buchanan and Whiting (1996) published a three-stage risk assessment example concerning *Salmonella* spp. in cooked chicken. This early study was not intended to follow the *Codex Alimentarius* approach, but was an illustration of the use of predictive microbiology. The production process is that raw chicken is stored at 10°C for 48 hours before being cooked at 60°C for 3 minutes and is then stored at 10°C for 72 hours before consumption. The 10°C stored temperature is in the 'danger zone' of microbial growth and represents mild temperature abuse.

Stage 1. Number of Salmonella *spp. in raw chicken before cooking*
The number of *Salmonella* cells on raw chicken will vary; however, an expected level of contamination is given in Fig. 4.5. The contamination

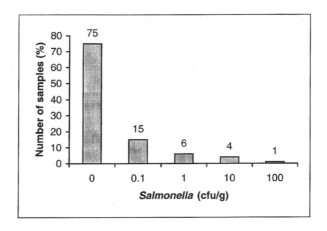

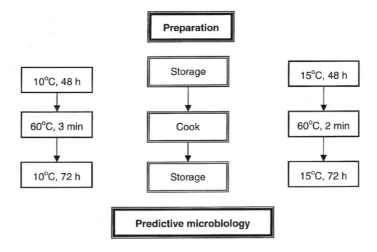

Salmonella (%)									
Safe process					Temperature abuse				
75	15	6	4	1	75	15	6	4	1
P_i 0	8.8×10^{-11}	6.5×10^{-10}	5.1×10^{-5}	4.1×10^{-8}	0	1.1×10^{-1}	4.1×10^{-1}	7.0×10^{-1}	8.6×10^{-1}

Fig. 4.5 Probability of *Salmonella* infection per gram of cooked chicken after temperature abuse (Buchanan & Whiting 1996).

range varies from no *Salmonella* cells in 75% of samples to 1% containing 100 cells per gram of meat. The amount of *Salmonella* growth at 10°C for 48 hours before cooking can be determined using growth models (Section 2.7) by assuming the meat is of pH 7.0 and the sodium chloride level is 0.5%.

Stage 2. Effect of cooking (60°C, 3 minutes) on Salmonella *numbers in chicken*
The decimal reduction time (*D* value) at 60°C is 0.4 minutes. The effect of heat treatment on *Salmonella* numbers can be calculated using the equation

$$\log(N) = \log(N_0) - (t/D)$$

where *N* is the number of micro-organisms (cfu g^{-1}) after the heat treatment, N_0 is the initial number of bacteria (cfu g^{-1}), *D* is the decimal reduction time (log(cfu g^{-1}) min^{-1}), and *t* is the duration of the heat treatment (min). Note that, for simplicity, no effect on *Salmonella* numbers is taken into consideration for the time-period during warming the food to 60°C and cooling afterwards. This equation gives the number of surviving *Salmonella* after the cooking process and is designed to give a 7*D* kill.

Stage 3. Salmonella *cell numbers following storage at 10°C, 72 hours before consumption*
As before, in stage 1 the growth curve for *Salmonella* can be predicted to estimate the number of organisms in cooked chicken after storage but before consumption.

By determining the survival number and subsequent growth for each initial population level of *Salmonella*, an estimate is given of the numbers of *Salmonella* that a population of consumers is likely to ingest. In this example, 1% of the chicken samples contained 100 *Salmonella* cells per gram, giving a probability of infection (P_i) of 4.1×10^{-8} per gram of food consumed (see Section 3.3.6 for an explanation of P_i). This means there was less than one cell surviving for every 10 000 g (10 kg) of food. Hence the *Salmonella* risk associated with cooked chicken under these conditions of storage is minimal.

The above example can be used as a template to determine the effect of changing the cooking regime and storage conditions. For example, raising the initial storage temperature to 15°C and reducing the cooking time to 2 minutes causes P_i to be unacceptably high (Fig. 4.5, right hand side).

4.2.6   Salmonella *spp. in cooked patty*

Whiting (1997) published an early microbiological risk assessment for *Salmonella* spp. in cooked patty that was in five steps.

(1)   Initial distribution
   The initial microbial load was taken from published data (Surkiewicz *et al.* 1969), where 3.5% of the samples have *Salmonella* present at levels greater than 0.44 cfu g^{-1} (Fig. 4.6).
(2)   Storage
   Conditions chosen were 21°C for 5 hours and the growth rate was predicted from a *Salmonella* model (Gibson *et al.* 1988).
(3)   Cooking
   Published *D* values were used to determine the effect of heat treatment at 60°C for 6 minutes.
(4)   Consumption
   A typical serving of 100 gram was assumed.
(5)   Infectious dose determination.

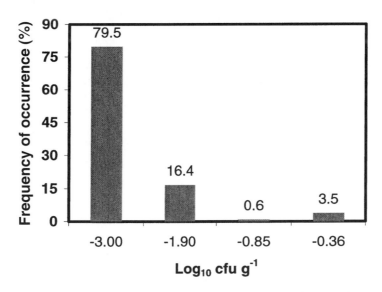

Fig. 4.6   Frequency of *Salmonella* in poultry patty (Whiting 1997).

The model calculated that one *Salmonella* cell has a mean probability of $10^{-4.6}$ of being an infectious dose (Table 4.3). However, 3% of the predictions gave risks greater than 10^{-3} because of a small number of initial samples with high *Salmonella* contamination. The usefulness of the model is in determining the effect of altering the variables, such as the

Table 4.3 Risk assessment model for *Salmonella* in a cooked poultry patty (adapted from Whiting 1997).

Stage	Statistics
Initial distribution	-2.7 log cfu g^{-1}
Storage (21°C for 5 hours)	0.17 log cfu g^{-1}
Cooking (60°C for 6 minutes)	-4.42 log cfu g^{-1}
Consumption (100 g)	-2.42 log cfu g^{-1}
Infectious dose (probability)	$10^{-4.6}$

cooking temperature to 61°C, which reduces the median probability of $10^{-7.4}$.

The approach of Whiting (1997) is very similar to that of Miller *et al.* (1997) for *L. monocytogenes* (Section 4.4.2).

4.2.7   Poultry FARM

A series of predictive models for *Salmonella* and *C. jejuni* infections from chicken called 'Poultry Food Assess Risk Model' (Poultry FARM) can be downloaded from the Internet (see Internet Directory). The models use @RISK™ (Palisade Corp.) and Excel™ (Microsoft) spreadsheets to predict the change in microbial load of 100 000 servings of chicken between packaging and processing. Additionally, the model predicts the number of cases of severe outcomes per 100 000 servings and the overall public health impact of the chicken. The Internet site includes a full (58 pages long) explanation of the Poultry FARM model.

4.2.8   Domestic and sporadic human salmonellosis

Hald *et al.* (2001) produced an alternative quantitative risk assessment method using a 'Bayesian Monte Carlo' approach which combined Bayesian inference with Monte Carlo simulation. The method was applied to quantifying the contribution of animals to domestic and sporadic human salmonellosis. Data from the 1999 Danish national surveillance of *Salmonella* in animals, foods and humans was used to demonstrate the method as an alternative to 'stable to table'.

The number of domestic and sporadic cases (caused by different *Salmonella* serovars) was estimated from the registered number of cases.

This was then compared with the prevalence of the *Salmonella* serovars isolated from different animal sources, weighted according to the consumption pattern and the association of particular *Salmonella* serovars with specific food sources. The probability of observing the actual number of human cases was determined using a Poisson likelihood function from the data of *Salmonella* prevalence in the various food types and amount ingested. The formula used was

$$\lambda_{ij} = M_j * p_{ij} * q_i * a_j * (\text{non-se})$$

where λ_{ij} is the expected number of cases per year of type i from source j, M_j is the amount of source j available for consumption per year, p_i is the prevalence of type i in source j, q_i is the bacteria-dependent factor for type i, a_j is the food-source-dependent factor for source j, and non-se means the serotype is not *S.* Enteritidis and the source is eggs; otherwise non-se is equal to unity.

The Bayesian Monte Carlo technique was then used to determine the distribution of cases with food sources. The most important source was estimated to be eggs (54%), pork (9%) and poultry (8%). A fuller mathematical account is given in the original article (Hald *et al.* 2001).

4.3   *Campylobacter jejuni* and *C. coli*

Sources:

- Livestock: pigs, cattle and sheep
- Domestic animals: cats and dogs
- Poultry
- Raw milk
- Polluted water.

Control measures:

- Heat treatment (pasteurisation/sterilisation)
- Hygienic slaughter and processing procedures
- Prevention of cross-contamination
- Good personal hygiene
- Water treatment.

Campylobacter are Gram-negative microaerophilic bacteria. However, their tolerance to oxygen is strain- and species-dependent. There are two major species of *Campylobacter* causing food poisoning: *C. jejuni* causes the majority of outbreaks (89–93%) and *C. coli* causes 7–10% of cases, whereas *C. upsaliensis* and *C. lari* have only occasionally been implicated. The reservoirs of *Campylobacter* spp. include poultry, cattle, swine, sheep, rodents and birds (Skirrow 1991). The routes of infection are via contaminated water, milk and meat (see Table 1.2). Poultry is the largest potential source of infectious *Campylobacter* organisms. Consequently most sporadic infections are associated with improper preparation of poultry or consumption of mishandled poultry products. Most *C. jejuni* outbreaks, which are far less common than sporadic illnesses, are associated with the consumption of raw milk or unchlorinated water. Direct enumeration of *Campylobacter* species is rarely possible. Usually an enrichment step is used to recover low numbers of the organism from processed foods.

The organism does not multiply at room temperature (minimum growth temperature is 30°C; Table 2.4). Therefore *C. jejuni* and *C. coli* do not multiply in chilled foods, although they will persist under such conditions. However, its reportedly low infectious dose of 500 cells (see Section 3.3.6 for clarification) means it can easily cause cross-contamination from raw meats to processed meats. This may be why cases of *Campylobacter* gastroenteritis outnumber salmonellosis in many countries. There is a notable seasonality to *Campylobacter* enteritis, with peak incidences in the summer months. Temperature, pH and water activity growth ranges and *D* values for *C. jejuni* are given in Table 2.4. Under stress conditions, *C. jejuni* may change into the so-called 'viable but non-culturable' (VNC) form and hence not be detectable by standard laboratory techniques.

An increase in resistance to ciprofloxacin (a medically important antibiotic) has been reported, possibly resulting from the veterinary use of the structurally related (fluoroquinolone) antibiotic enrofloxacin used in poultry husbandary (see Section 1.3).

The characteristics of *Campylobacter* enteritis are the following:

- Flu-like illness
- Abdominal pain
- Fever
- Diarrhoea, may be profuse, watery and frequent or alternatively bloody.

The incubation period is 2–10 days, the disease lasts for about 1 week and is usually self-limiting. In 20% of cases the symptoms may last from 1 to 3 weeks. The organism is excreted in the faeces for several weeks after the symptoms have ceased. Relapses occur in about 25% of cases. There is no consensus of opinion concerning the virulence factors of *Campylobacter*. *Campylobacter* infection is now recognised as the single most identifiable antecedent infection associated with the development of Guillain–Barré syndrome (GBS).

Since the eradication of polio in most parts of the world, GBS has become the most common cause of acute flaccid paralysis (Allos 1998). GBS is an autoimmune disorder of the peripheral nervous system characterised by weakness, which is usually symmetrical, evolving over a period of several days or more. It occurs world-wide and is the most common cause of neuromuscular paralysis. *Campylobacter* enteritis is associated with several pathogenic forms of GBS, including the demyelinating (acute inflammatory demyelinating polyneuropathy) and axonal (acute motel axonal neuropathy) forms. Of an estimated annual number of 2628–9575 GBS cases in the USA, 526–3830 were triggered by *Campylobacter* infection. Some sources state that half the GBS cases may be due to *Campylobacter* infections. Typically, the gastrointestinal symptoms occur 1–3 weeks before the neurological symptoms. Different strains of *Campylobacter*, together with host factors, are likely to play an important role in determining who develops GBS as well as the peripheral nerves targeted by the host's immune system. GBS is probably due to an autoimmune response to the ganglioside GM1 on peripheral nerves following infection by *C. jejuni* O19 (although other serotypes may be involved), because peripheral nerves may share epitopes with surface antigens of *C. jejuni* (molecular mimicry). Cytokines may induce the inflammatory process and lead to nerve demyelination. Additionally, complement has a role in the process leading to nerve damage, which leads to an increase in the permeability of the blood–nerve barrier and hence causes the inflammation.

In addition to the following risk assessment examples, a dose–response curve is given in Chapter 3 (Fig. 3.9 and Table 3.9) and the FARM model is given in Section 4.2.7.

4.3.1   C. jejuni *risk from fresh chicken*

Fazil *et al.* (2000b) have produced a quantitative model for the risk from *C. jejuni* on fresh chicken which can be accessed from the Web (see Internet Directory). The model determines the fate of the organism through the food chain ('farm-to-fork') using a slightly different risk assessment framework: hazard identification, exposure assessment, dose–response and risk characterisation. This was termed the 'process risk model' (see also *E. coli* O157:H7, Section 4.5). The objectives of the study were to generate a model of the production of chickens, gain a better understanding of chicken processing operations and identify important steps in the process that influence risk. Hence the study would generate a 'tool' to assist decision making in order to reduce the risk to the consumer from food-borne pathogens. The study did not include long-term illnesses such as GBS.

Figure 4.7 summarises the food chain for *C. jejuni* in chicken and includes the variation in data obtained from published literature; it can be compared with the framework model of Fig. 3.8. The *Campylobacter* prevalence going into the process (P_F) was also described using a beta distribution (mode = 56%). The effect of processing on *C. jejuni* prevalence and concentration was determined for the following five stages: scalding, defeathering, evisceration, washing and chilling. Due to considerable uncertainty in the literature, the numbers of *C. jejuni* could either increase or decrease. Due to the absence of complete information a triangular distribution (Section 3.3.9, Fig. 3.11) was used which gave minimum, maximum and most likely parameters. The well-studied physiology of the organism (Table 2.4) led to the assumption that after processing no multiplication would occur during transit from the 'processing plant' to the home. It has previously been determined that 20–30% of meals are undercooked. In addition, given the temperature sensitivity of the organism (Table 2.4), it was assumed that 30–40% of *C. jejuni* on contaminated chickens would be in areas that were protected from direct heat. Previously published *D* values were used to determine the effect of cooking on the count of viable *C. jejuni*. These calculations subsequently generated the probability of exposure and the dose likely to be ingested by the consumer. Using dose–response analysis, the probability that the consumer will be infected at the given ingested dose can be determined. The dose–response analysis of Medema *et al.* (1996) using human feeding trial data from Black *et al.* (1988) has previously been given in Fig. 3.9 (see Table 3.9 for original data).

The prevalence of contaminated carcasses leaving the processing plant and the concentration of *C. jejuni* on the carcasses was determined by Fazil *et al.* (2000b) using combined @RISK™ (Palisade Corp.)-Microsoft Excel™ (Microsoft Corp.) software to incorporate the Monte Carlo analysis. The model was run for 10 000 iterations. The data were also used to determine the probability of illness and number of illnesses. The variation in values at each stage illustrates the need for mathematical simulations to determine the variation and uncertainty of the estimates. (See Section 3.3.10 for an explanation of Monte Carlo analysis and related topics.) The most likely prevalence of *C. jejuni* on chickens was 65–85%, with an average of $\log_{10} 3.8$ (i.e. 6310) *C. jejuni* cells per carcass. The distribution of probability of infection per serving of one-quarter of a chicken was 2.23×10^{-4}. These values can be converted into an estimate of number of illnesses per year (in the USA) by using the risk per serving, the number of chicken servings consumed in a year and the size of the population consuming the chickens. The predicted number of illnesses is in the order of

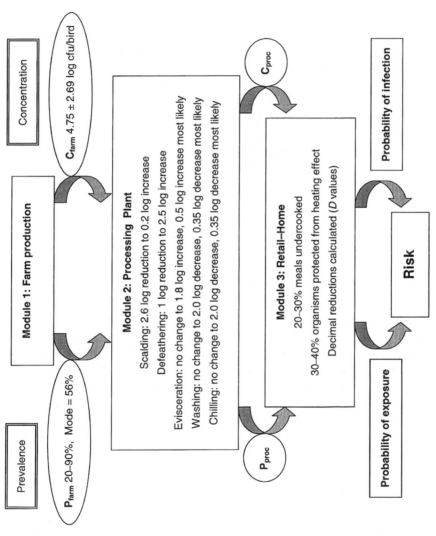

Fig. 4.7 *Campylobacter jejuni* in chicken (adapted from Fazil *et al.* (2000b); see also Fig. 3.3).

2.7 million per year. The microbiological risk assessment also predicts the parameters ('importance analysis') that are important in influencing the risk (tornado chart, Fig. 4.8) either because of their uncertainty or variability. Uncertain parameters require future research, whereas variable factors could be important control points by which to reduce the risk.

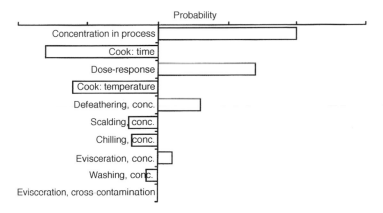

Fig. 4.8 *Campylobacter jejuni* importance analysis.

4.3.2 Risk profile for pathogenic species of Campylobacter in Denmark

In 1998 the Danish Veterinary and Food Administration started to collate a risk profile for *C. jejuni*. Afterwards the study was used to produce a risk assessment of human illness from *C. jejuni* in broilers (Christensen *et al.* 2001; Section 4.3.3). The studies collated information from a wide range of studies within Denmark and have been summarised below and in Section 4.3.3 to demonstrate a country-specific microbiological risk assessment. Part of the reasoning for this work is the general trend of governments to reduce the prevalence of pathogens such as *Campylobacter* in poultry. The Dutch government has announced that the prevalence of *Campylobacter* in poultry has to be reduced to 0%, i.e. completely eliminated. However, the Product Board for Livestock, Meat and Eggs in The Netherlands can only suggest a plan for reducing the occurrence of *Campylobacter* in poultry meat to below 15% of carcasses.

The rate of incidence of *Campylobacter* gastroenteritis in Denmark was approximately 50 reported cases/100 000 population, although it was estimated that the actual number of cases was approximately twenty times greater due to under-reporting (Christensen *et al.* 2001). In Denmark, the number of sporadic cases peaks in the summer months, whereas the number of outbreaks culminates in May and October. The cases were more prevalent in the 10–19 year age group. A case-control investigation of food-borne risk factors for sporadic campylobacteriosis in Denmark (May 1996 to September 1997) used 227 cases and 250 control persons. On the basis of the established risk factors, the aetiology of approximately 50% of the human cases could be determined: 5–8% were due to consumption of insufficiently cooked poultry, 15–20% to consumption of grilled meat, 5–8% to contaminated potable water and 15–20% to overseas travel.

The incidence of *Campylobacter* spp. in farm animals was studied for pigs, cattle, broilers and turkeys by Christensen *et al.* (2001). *C. coli* and *C. jejuni* were detected in 95% and 0.3%, respectively, of pig slurry samples. It has also been shown that before freezing 66% of pig carcasses contained *Campylobacter* whereas after freezing only 14% did so. The prevalence of *Campylobacter* in cattle was 51%, with *C. jejuni* as the most frequently isolated species. *Campylobacter* prevalence in broilers was 37%, with a distinct seasonal variation, the most frequent incidence being in the summer. *C. jejuni* was the most frequently isolated species. The incidence in turkeys was assumed to be similar to that in chickens. *Campylobacter* was much more likely to be present in fresh poultry (20–30% of carcasses affected) than in beef or pork (1% of samples affected).

Human exposure to *Campylobacter* is not only from food, but from the environment (such as bathing water), wild animals and pets. *Campylobacter* was detected in 29% of faeces samples from dogs less than 5 months old, 16 of the 21 isolates being *C. jejuni*, while *C. upsaliensis* was detected in two out of 42 faeces samples from cats less than 5 months old.

Ongoing research programmes include determining *Campylobacter* prevalence in environmental reservoirs (such as private wells and bathing water) and unpasteurised milk, the incidence and significance of the 'viable but non-culturable form', dose–response relations, the exact incidence in the population, and the significance of chronic sequelae. The study by Christensen *et al.* (2001) provided data to initiate a risk assessment of *Campylobacter* from broiler chickens outlined below.

4.3.3 Risk assessment of C. jejuni *in broilers*

As described above, the Danish Veterinary and Food Administration produced a risk profile of *C. jejuni* which was subsequently used to help produce a risk assessment of *C. jejuni* in broilers (Christensen *et al.* 2001). The prevalence of *C. jejuni* in a flock and their concentration after bleeding was the starting point of the assessment model which had two subsequent modules: (1) slaughter and processing, and (2) preparation and consumption. Cross-contamination between contaminated and uncontaminated flocks was modelled, but cross-contamination within flocks was not.

The prevalence and concentration of *C. jejuni* in each module was estimated and used to describe a dose–response relationship with associated probability of infection and probability of illness. The probability of illness associated with chilled chickens and frozen chickens was 1 case per 6000 meals and 1 case per 26 000 meals, respectively. The probability of illness was comparable with the (1999) registered number of *Campylobacter* enteritis cases in Denmark (Fig. 4.9).

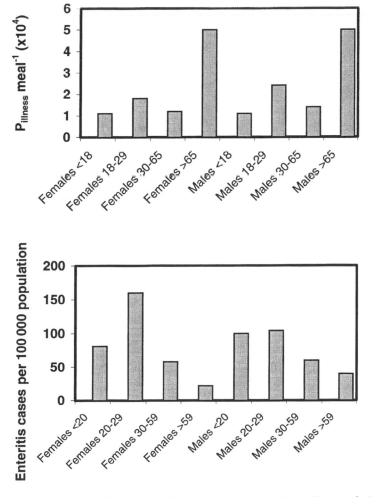

Fig. 4.9   Probability of illness and *Campylobacter* enteritis in Denmark (Christensen *et al.* 2001).

As before, risk assessment enables predictions on the effect of risk mitigation strategies. In this case a 25-fold reduction in the number of human illnesses could be obtained by the following:

(1) A 100-fold reduction in the concentration of *C. jejuni* on the broiler chickens (i.e. from $1000\,\text{cfu}\,\text{g}^{-1}$ to $10\,\text{cfu}\,\text{g}^{-1}$)
(2) A 25-fold reduction in the flock prevalence (i.e. from 60% to 2.4%)
(3) A 25-fold increase in 'safe' consumer behaviour during preparation of a chicken meal.

4.3.4   Campylobacter *fluoroquinolone resistance*

Approximately one-half of the antimicrobials produced today are used in human medicine. Most of the remainder are added to feed for food animals, either for mass treatment against infectious diseases or to promote the growth of pigs and poultry. The increasingly widespread use of antimicrobials outside human medicine has become the subject of particular concern, with the alarming emergence in humans of bacteria which have acquired resistance to medically important antibiotics (FDA 1998).

A significant proportion of the rising antimicrobial resistance problem in human medicine is due to the overuse and misuse of medically important antimicrobials. However, some of the newly emerging antibiotic resistant bacteria are transmitted to humans mainly via meat and other foods of animal origin, or by direct contact with farm animals. This has been reported for the food-borne bacteria *Salmonella* and *Campylobacter* and the commensal bacterium *Enterococcus*. The rise of ciprofloxacin (a medically important fluoroquinolone) resistance is linked to the use of the veterinary antibiotic enrofloxacin which is also a fluoroquinolone. The mechanism of resistance is through the mutation of DNA gyrase which confers cross-resistance to ciprofloxacin. It is plausible that the prolonged use of enrofloxacin has exerted a selective pressure for spontaneous mutations of DNA gyrase, the target enzyme of both antibiotics.

The US Center for Veterinary Medicine (FDA 2000a) has produced a risk assessment on the human health impact of fluoroquinolone resistant *Campylobacter* associated with the consumption of chicken. This is available on the Web (see Internet Directory, FDA). This risk assessment also includes two models, one which uses Excel™ (Microsoft Corp.) and is hence readily available, and the other which uses @Risk (Palisade Corp.) and can be modified for Crystal Ball users (Decisioneering). The risk values obtained are related to 1998 input data only.

The model is summarised in Fig. 4.10. It is divided into five modules:

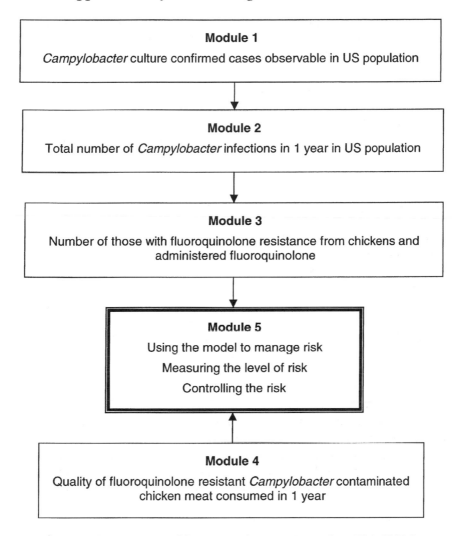

Fig. 4.10 Risk assessment of fluoroquinolone use in poultry (FDA 2000a).

(1) *Campylobacter* culture confirmed cases observable in the US population
(2) Total number of *Campylobacter* infections in 1 year in the US population
(3) Number of those with fluoroquinolone resistance from chickens and administered fluoroquinolone
(4) Quantity of fluoroquinolone resistant *Campylobacter* contaminated chicken meat consumed in 1 year
(5) (i)   Using the model to manage risk.

(ii)   Measuring the level of risk.
(iii) Controlling the risk.

The reader should note that for clarity the author has used the term 'module' whereas the original document uses the term 'section'. Table 4.4 summarises the results.

Table 4.4   Risk assessment on the human health impact of fluoroquinolone (FDA 2000a).

Module	Step	Value		
1	US population	270 298 524		
	Catchment site population	20 723 982		
	Observed FoodNet invasive cases	43		
	Observed FoodNet enteric cases	3 985		
	Estimated mean population invasive infections	5 621		
	Estimated mean population enteric infections	51 976		
	Estimated mean culture confirmed cases		Enteric	Invasive
		Non-bloody	Bloody	
		28 077	23 898	561
2	Proportion seeking care	12%	26.7%	100%
	Proportion submitting stool sample	19%	55.4%	100%
	Proportion of samples tested in laboratory	94.5%	94.5%	100%
	Portion of cultures confirmed	75%	75%	100%
	Illness in population	1 702 043	228 040	561
	Total number of cases		1 930 644	
3	Number of fluoroquinolone resistant infections from chickens (59%)	1 004 205	134 543	331
	Proportion seeking care	12.2%	26.7%	100%
	Number seeking care	122 078	35 878	331
	Proportion treated with antibiotic	47.9%	63.7%	100%
	Number treated	58 450	22 854	331
	Proportion receiving fluoroquinolone treatment	55.08%	55.08%	55.08%

Contd

Table 4.4   *(Contd)*

Module	Step	Value		
	Number of chicken related cases treated with fluoroquinolone	32 195	12 588	182
	Proportion of *Campylobacter* infections from chicken that are fluoroquinolone resistant	10.4%		
	Number of fluoroquinolone resistant infections from chicken seeking care, receiving fluoroquinolone	3 352	1 311	19
	Total number of fluoroquinolone resistant infections from chicken seeking care, receiving fluoroquinolone		4 682	
4	Total prevalence of *Campylobacter*		88.1%	
	Prevalence of fluoroquinolone resistant *Campylobacter* among *Campylobacter* isolates from slaughter plant		11.8%	
	Estimated prevalence of fluoroquinolone resistant *Campylobacter* in broiler carcasses		10.4%	
	Chicken consumption per head	51.4 lb (23.3 kg)		
	Total consumption in USA	1.39×10^{10} lb (6.3×10^{6} tonnes)		
	Total consumption of chicken with fluoroquinolone resistant *Campylobacter* in USA	1.45×10^{9} lb (6.57×10^{5} tonnes)		

Module 1 explains the process of extrapolating the number of culture-confirmed cases reported to the US Communicable Disease Surveillance Center, to the total number of culture-confirmed cases in the USA, and subdividing the number according to whether the infection is invasive or enteric (with and without bloody diarrhoea).

Module 2 uses the values calculated in Module 1 to estimate the predicted total number of *Campylobacter* cases in the USA. This gives a mean estimate of 1.92 million cases per year (1.6–2.6 million, 90% confidence interval).

Module 3 shows that about 5000 people (2585–8595, 90% confidence interval), who were in fact infected with fluoroquinolone resistant *C. jejuni* from consuming contaminated chicken, were given fluoroquinolone treatment.

Module 4 estimates that 1450 million lb (about 658 000 tonnes) (967 million to 2000 million lb (438 000–907 000 tonnes), 90% confidence interval) of boneless chicken contaminated with fluoroquinolone resistant *C. jejuni* was consumed.

Module 5 determines the final estimates of risk. For the average US person, only one in 61 093 were affected by the use of fluoroquinolone, whereas for the target subpopulation (person with *Campylobacter* enteritis seeking care and prescribed fluoroquinolone) the risk increased to one in 30.

The model can be used to determine the maximum prevalence of fluoroquinolone resistant *C. jejuni* that is permissible before there is an unacceptable human health impact.

4.4   *Listeria monocytogenes*

Sources:

- Soil and vegetables
- Wild and domestic animals: meat and milk
- Infected humans
- Sewage.

Control measures:

- Heat treatment of milk (pasteurisation, sterilisation)
- Avoidance of cross-contamination
- Refrigeration (limited period) followed by thorough reheating
- Avoidance of high risk products (e.g. raw milk) by high risk populations (e.g. pregnant women).

Listeria are Gram-positive, non-spore-forming bacteria. They are motile by means of flagella and grow between 0 and 42°C. They are more sensitive to heat than *Salmonella* and hence pasteurisation is sufficient to kill the organism. The genus is divided into eight species, of which *L. monocytogenes* is the species of primary concern with regard to food poisoning. The species is further subdivided using serotyping. The epidemiologically important serotypes are 1/2a (15–25% of cases), 1/2b (10–35% of cases) and 4b (37–64% of cases). Outbreaks due to serotype 4b are significantly greater in pregnancy cases, whereas serovar 1/2b is more common with non-pregnancy cases (McLauchlin 1990a,b; Farber & Peterkin 1991). Infection in healthy adults is typically asymptomatic and it is plausible that 1–10% of humans may be intestinal carriers of *L. mono-cytogenes*. The majority of those who succumb to severe listeriosis are individuals with underlying conditions that suppress their T-cell mediated immunity. Another factor related to susceptibility is reduced gastric acidity as occurs on ageing, especially in those more than 50 years old. This may explain the age distribution of listeriosis (Fig. 4.11).

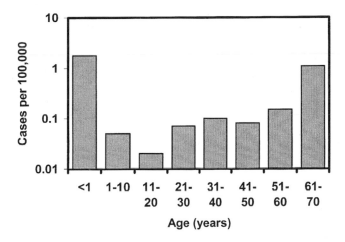

Fig. 4.11   Distribution of listeriosis with age (Buchanan & Linqvist 2000).

This ubiquitous bacterium has been isolated from various environments, including decaying vegetation, soil, animal feed, sewage and water. It is found in a wide variety of foods, both raw and processed, where it can survive and multiply rapidly during storage (See Table 1.2). These foods include supposedly pasteurised milk and cheese (particularly soft-ripened varieties), meat (including poultry) and meat products, raw vegetables, fermented raw-meat sausages as well as seafood and fish products. *L. monocytogenes* is quite hardy and resists the deleterious effects of freezing, drying and heat remarkably well for a bacterium that does not form spores. Its ability to grow at temperatures below 8°C permits its multiplication in refrigerated foods.

L. monocytogenes is responsible for opportunistic infections, preferentially affecting individuals whose immune system is perturbed, including pregnant women, newborns and the elderly. Listeriosis is clinically defined when the organism is isolated from blood, cerebrospinal fluid, or an otherwise sterile site, e.g. placenta or foetus.

Symptoms of listeriosis are the following:

- Meningitis, encephalitis or septicaemia
- When pregnant women are infected in the second or third trimesters, abortion, stillbirth or premature birth can result.

The infective dose of *L. monocytogenes* is unknown, but is believed to vary with the strain and susceptibility of the victim. The CCFH has stated that a concentration of *L. monocytogenes* not exceeding 100 organisms per gram of food at the point of consumption is of low risk to the consumers (CCFH 1999).

From cases contracted through raw or supposedly pasteurised milk it is evident that fewer than 1000 organisms may cause disease. The stomach acidity will reduce the viable number of cells, but it requires exposure times of between 15 to 30 minutes to reduce *L. monocytogenes* numbers by 5-log orders. Additionally, ingested small liquid volumes of less than 50 ml may pass rapidly through the stomach because the pyloric sphincter will not be stimulated to contract. Another factor affecting survival in the stomach is the food matrix, especially the presence of fatty material (see Fig. 3.7). Cells that survive passage through the stomach can invade sites along the intestinal tract, in particular the small intestine, either through the M cells overlaying the Peyer's patches or through non-specialised enterocytes. Once the bacterium has entered the host's monocytes, macrophages or polymorphonuclear leukocytes it is blood-borne (septicaemic) and can grow. Its intracellular presence in phagocytic cells also permits access to the brain and probably transplacental migration to the foetus in pregnant women. The pathogenesis of *L. monocytogenes* centres on its ability to survive and multiply in phagocytic host cells. The incubation period is extremely variable at 1–90 days. Listeriosis has a very high mortality rate. When listeric meningitis occurs, the overall mortality may be as high as 70%. Cases of septicaemia have a 50% fatality rate, whereas the fatality rate for perinatal–neonatal infections is greater than 80%. The mother usually survives infections during pregnancy. Infection can be symptomless, resulting in faecal excretors of infectious *Listeria*. Consequently, about 1% of faecal samples and about 94% of sewage samples are positive for *L. monocytogenes*.

4.4.1   L. monocytogenes *hazard identification and hazard characterisation in ready-to-eat foods*

JEMRA (Buchanan & Lindqvist 2000) published a preliminary report on the hazard identification and hazard characterisation of *L. monocytogenes* in ready-to-eat (RTE) foods as part of the on-going programme of microbiological risk assessments. The objectives were to quantitatively evaluate the nature of the adverse health effects associated with *L. monocytogenes* in RTE foods, and to assess the relationship between magnitude of the dose and the frequency of these health effects. The entire document prepared by Buchanan & Lindqvist (2000) was part of the JEMRA 2000 meeting (Table 3.2) and can be downloaded (see Internet Directory). RTE foods are considered by the *Codex Alimentarius* to be any food or beverage that is normally consumed in its raw state, or any that is handled, processed, mixed, cooked, or otherwise prepared into a form in which it is normally consumed without further processing.

The report has an extensive literature survey and various dose–response models. Due to the lack of human feeding trials and surrogate pathogens to determine the probability of infection (Section 3.3.6), data from epidemiological studies and animal models were used. Various dose–response models were assessed which had different end-points (infection, morbidity and mortality). The dose–response relationship which best described the interaction between *L. monocytogenes* and humans was not resolved, though the highly variable response is known to be dependent upon the combined interaction of the host, pathogen and food matrix. Subsequently, it was recommended that several multiple dose–response models should be used in developing the risk assessment. Data from animal studies that modelled lethality or severe invasive listeriosis was more related to human disease than modelling infection (Section 3.3.6).

A summary of selected dose–response models that were reviewed is given in Table 4.5, and Figs 4.12 and 4.13 (murine model) show the dose–response models for frequency of infection (Gompertz-log) and mortality (exponential). The mathematical models are described in Section 3.3.6. A comparison of the FDA models for general population, neonates and elderly with dose–response curves from only epidemiological data showed a lower median probability of response at a specified dose (Fig. 4.12). The difference was probably due to the FDA models being based on mortality not morbidity, and that other models were based on the use of highly virulent strains of *L. monocytogenes*. The predicted risk of serious listeriosis was determined to be five times the risk of mortality.

4.4.2 L. monocytogenes *exposure assessment in ready-to-eat foods*

JEMRA (Ross *et al.* 2000) published a preliminary report (which was subsequently withdrawn pending correction to the *Listeria* risk assessment simulation model) on the exposure assessment for *L. monocytogenes* in RTE foods as part of the on-going programme of microbiological risk assessments (Section 3.2.4). The report both reviewed (extensively) the relevant current publications and further developed an exposure assessment for *L. monocytogenes*. In addition to

Table 4.5 Dose–response models for *L. monocytogenes* (Buchanan & Lindqvist 2000).

Model/Study	Biological end-point	Model/Parameters	Comments
Buchanan *et al.* (1997)	Serious listeriosis Based on annual statistics and food survey data	Exponential $P_i = 1.18 \times 10^{-10}$	Based on immunocompromised individuals Predicted morbidity$_{50} = 5.9 \times 10^9$ cfu
Lindqvist & Westöö (2000)	Serious listeriosis Based on annual statistics and food survey data	Exponential $P_i = 5.6 \times 10^{-10}$	Based on immunocompromised individuals Predicted morbidity$_{50} = 1.2 \times 10^9$ cfu
Chocolate milk Buchanan & Linqvist (2000)	Febrile gastroenteritis Outbreak data	Exponential $P_i = 5.8 \times 10^{-8}$	Based on chocolate milk outbreak and limited to immunocompromised individuals
Farber *et al.* (1996)	Serious infection	Weibull-gamma = 0.25 high risk = 10^{11} $b = 2.14$	Estimated dose for 50% population to be infected: High risk = 4.8×10^5 cfu Low risk = 4.8×10^7 cfu

(Contd)

Table 4.5 *Contd*

Model/Study	Biological end-point	Model/Parameters	Comments
Butter Buchanan & Lindqvist (2000) FDA (2001)	Serious listeriosis Outbreak data	Exponential $P_i = 1.02 \times 10^{-5}$	Outbreak data from Finland for immunocompromised individuals Predicted morbidity$_{50}$ = 6.8 × 10^4 cfu
Mexican-style cheese Buchanan & Lindqvist (2000) FDA (2001)	Morbidity Outbreak data	Exponential $P_i = 3.7 \times 10^{-7}$	Outbreak data from USA Predicted morbidity$_{50}$ = 1.9 × 10^6 cfu
FDA-General FDA-Neonates FDA-Elderly FDA (2001)	Mortality Combined human and surrogate animal (murine) data (Golnazarian *et al.* 1989)	Five models compared At 10^{12} dose: P_i (general) = 8.5 × 10^{-16} P_i (neonates) = 5.0 × 10^{-14} P_i (elderly) = 8.4 × 10^{-15}	Gompertz-log equation was best fit for frequency of infection, whereas exponential model was best for mortality

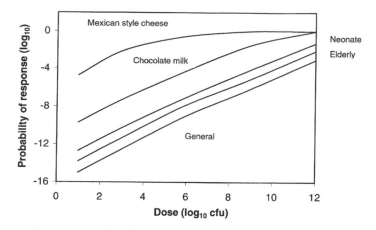

Fig. 4.12 *Listeria monocytogenes* dose–response curves for various age groups and sources (Buchanan & Lindqvist 2000).

reviewing eleven risk assessments, the report gave seven new examples of exposure assessments in RTE foods. Raw and unpasteurised milk, ice cream and soft mould-ripened cheese were modelled from retail to point of consumption. Minimally processed vegetables, smoked salmon and semi-fermented meats were modelled from production to point of consumption. The risk of 100 cfu *L. monocytogenes* g^{-1} at point of consumption was compared with the effect of the 'zero tolerance' policy. The aim of these examples was to illustrate the effect on exposure of processing, low contamination levels in products that do not permit growth

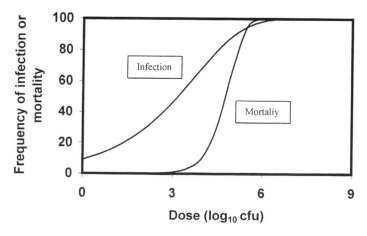

Fig. 4.13 Dose–response curves for illness and mortality following administration of *L. monocytogenes* to mice (Buchanan & Lindqvist 2000).

of *L. monocytogenes*, long-term storage on increase or decrease of *L. monocytogenes* concentration and, finally, consumption frequency and meal size. Additionally, the report covered the use of predictive microbiology in modelling risk assessments (Section 2.7).

A generic model was developed which emphasised the need to monitor changes in the prevalence and concentration of *L. monocytogenes* in RTE foods (Fig. 4.14). An example exposure assessment predicts that consumers at normal risk consume between 4 and 22 servings of soft cheese per year, and consumers at high risk consume between 3 and 17 servings per year. Of those servings, 4% (median) are predicted to be contaminated.

A problem with exposure assessment for *L. monocytogenes* (as well as for *Salmonella* serotypes) is that current data often lack information on the organism's concentration in food because many authorities have a 'zero tolerance' policy. Hence laboratory procedures are often designed to determine only the presence or absence of the organism in a 25 g sample. The lack of sufficient incidence and prevalence data at the point of consumption means that at present predictive microbiology models have to be used. These models will need to be further validated in products of similar microbial ecology and extrinsic parameters.

The 'zero tolerance' policy was not shown to provide a greater level of public health than other less stringent criteria such as ensuring these were less than 100 *L. monocytogenes* cells per gram of food.

4.4.3   Relative risk of **L. monocytogenes** *in selected ready-to-eat foods*

A draft risk assessment of the relative risk to public health from food-borne *L. monocytogenes* in selected RTE foods was released by the United States Department of Agriculture in January 2001 (FDA 2001) for further comment. The risk assessment was separated into three subpopulations: perinatal (foetuses and new-borns less than 30 days after birth), elderly and 'intermediate-age' which was the remaining population both healthy and susceptible. The risk assessment was divided into four categories according to CAC (1999). The dose–response relationship was based on a mouse model (see Section 4.4.1; Buchanan & Lindqvist 2000) with murine death as the end-point as opposed to infection. A dose of 10^9 *L. monocytogenes* was chosen to compare the responses of the three age groups. The model predicted that for every 100 million servings (each containing 10^9 *L. monocytogenes*), the most likely numbers of deaths were as follows:

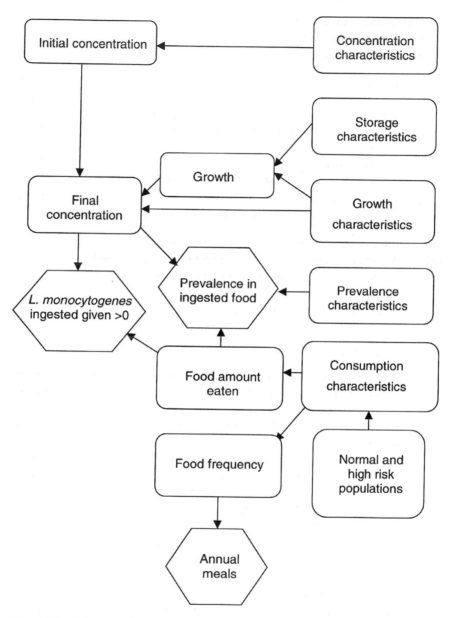

Fig. 4.14   Influence diagram for exposure assessment of *L. monocytogenes* in ready-to-eat foods (adapted from Ross *et al.* (2000)).

- Intermediate age-group, 103: range 1–1190
- Perinatal, 14 000: range 3125–781 250
- Elderly, 332: range 1–2350.

The considerable range was due to variability and uncertainty in the data used in the model. The risk characterisation used 300 Monte Carlo simulations (30 000 iterations per simulation). The risk characterisation also developed a 'relative risk ranking' for 20 food categories based on the predicted number of listeriosis cases per 100 million servings of the food category. The highest predicted risk RTE foods were pâté and meat spreads, smoked seafood, fresh soft cheese, frankfurters and some foods from delicatessen counters. The exposure assessment predicted that five factors affected consumer exposure to *L. monocytogenes*:

(1)   Amounts and frequency of food consumed
(2)   Frequency and level of *L. monocytogenes* in RTE food
(3)   Potential to support microbial growth during refrigeration
(4)   Refrigerated storage temperature
(5)   Duration of refrigerated storage before consumption.

4.4.4   L. monocytogenes *in EU trade*

The absence of agreed reference values for *L. monocytogenes* (except for dairy products) has led to controversy, especially concerning trading between member states of the EU. The lack of microbiological reference values has led to food-products being declared unfit for human consumption because of the non-quantified (zero tolerance) demonstration of *L. monocytogenes* contamination. Therefore microbiological risk assessment leading to the setting of food safety objectives for the EU was required and has been reported (Anon. 1999b).

Although human listeriosis is mainly caused by a few serovars (4b and 1/2a, b) it was concluded that a wide range of strains might cause serious disease. Additionally, because none of the typing methods discriminates pathogenic from non-pathogenic or less virulent strains, all *L. monocytogenes* were regarded as potentially pathogenic.

The EU risk assessment of *L. monocytogenes* enables six food groupings relative to processing control of the organism (Table 4.6). Examples of products are:

- Groups B and D: meat products such as cooked ham, wiener sausages or hot smoked fish, soft cheese made from pasteurised milk
- Groups C and E: cold smoked or gravid fish and meat, cheese made from unpasteurised milk
- Group F: tartar, sliced vegetables and sprouts.

Table 4.6 Grouping of ready-to-eat food commodities relative to the control potential for *L. monocytogenes*.

Group	Treatment
A	Foods heat-treated to a listericidal level in the final package
B	Heat-treated products that are handled after heat treatment. The products support growth of *L. monocytogenes* during the shelf life at the stipulated storage temperature
C	Lightly preserved products, not heat-treated. The products support growth of *L. monocytogenes* during the shelf life at the stipulated storage temperature
D	Heat-treated products that are handled after heat-treatment. The products are stabilised against growth of *L. monocytogenes* during the shelf life at the stipulated storage temperature
E	Lightly preserved products, not heat-treated. The products are stabilised against growth of *L. monocytogenes*, during the shelf life at the stipulated storage temperature
F	Raw, ready-to-eat foods

Groups B and D, and C and E are separated according to the technology used.

A concentration of 100 *L. monocytogenes* cells per gram of food at the point of consumption was considered a low risk to consumers. However, due to the uncertainties related to this risk, levels lower than 100 cells g^{-1} may be required for those foods in which *Listeria* growth may occur. *L. monocytogenes* levels above 100 cfu g^{-1} can be achieved after in-food growth. Therefore risk management should be focused on those foods which support *L. monocytogenes* growth.

Suggested levels of *L. monocytogenes* were as follows:

- Food groups D, E and F: < 100 cfu g^{-1} throughout the shelf-life and at point of consumption
- Food groups A, B, and C: not detectable in 25 g at time of production.

The food safety objective (Section 3.6) should be to keep the concentration of *L. monocytogenes* in food below 100 cfu g^{-1} and to reduce significantly the fraction of foods with a concentration above 100 *L. monocytogenes* cells per gram. In risk communication, special attention should be given to groups of consumers at increased risk (immunocompromised) which represent a considerable and growing section of the total population.

4.4.5 L. monocytogenes *in meat balls*

Miller *et al.* (1997) published a quantitative risk assessment for *L. mono-cytogenes* in meat balls to identify control points for HACCP implementation. A 'farm to fork' approach was used, as illustrated in Fig. 4.15. The risk assessment used data from the 1994 US census to identify the proportion of susceptible people in the general population (Table 4.7) and

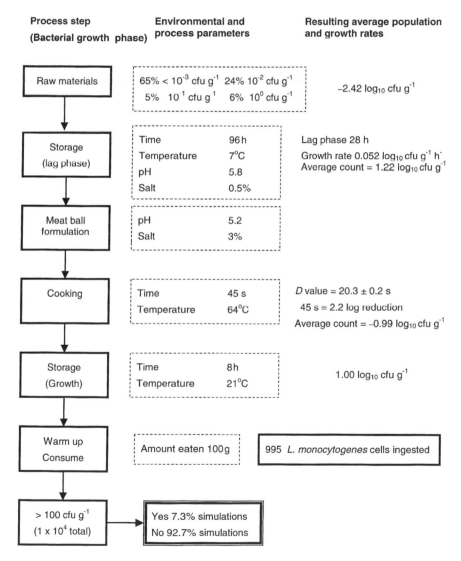

Fig. 4.15 Risk assessment of *L. monocytogenes* in meat balls (adapted from Miller *et al.* (1997)).

Table 4.7   US population distribution with respect to vulnerable groups (Miller *et al*. 1997).

Population category (year)	Percentage of US population
Pregnant women (1988)	2.5
Children < 5 years old (1992)	7.7
Elderly > 65 years (1992)	12.7
Residents in nursing and related care facilities (1991)	0.7
Cancer cases under care (1992)	1.6
Organ transplant patients (1992)	0.02
Total HIV (AIDS) infections (Jan. 1993)	0.03–0.04

combined the survey data with predictive microbiology (Pathogen Modeling Program, Section 2.7.3; see Internet Directory) and D values to simulate the changes in *L. monocytogenes* numbers (Table 2.4). A working morbidity threshold of about 100 *L. monocytogenes* cells per gram was used.

The risk assessment model estimated that eating 100 g of meatballs (processed as in Fig. 4.15) would result in the ingestion of an average of 995 *L. monocytogenes* cells. This was below the target of less than 100 cells g^{-1} (10^4 total dose) and hence appeared as a 'safe' process. However, there is considerable range in the prevalence and concentration of *L. monocytogenes* in the raw materials, and hence it is plausible that a high initial number of *L. monocytogenes* cells might be reduced to a dose greater than the target of 100 cells g^{-1}. Monte Carlo analysis (Section 3.3.10) was used to consider the frequency of samples that could exceed the target value. It was determined that 7.3% would exceed the target dose and hence the process was not as 'safe' as previous, more simple, estimations (Fig. 4.15).

The advantage of risk assessment models is that they can be used to predict the effect of processing changes and so forth on the risk. Miller *et al*. (1997) used the model to determine the effect of reducing the initial counts to 10^{-3} (71%), 10^{-2} (24%), 10^{-1} (5%) and 10^0 cfu g^{-1} (0%) (see distribution in Fig. 4.15). The model subsequently predicted that only 0.94% of the ingested doses would be greater than 100 cells g^{-1}.

4.5   Enterohaemorrhagic *E. coli* (EHEC); *E. coli* O157:H7

Escherichia coli is a Gram-negative, facultative anaerobe. The pathogenic strains are divided according to clinical symptoms and mechanisms of pathogenesis into the following groups:

- Enterohaemorrhagic *E. coli* (EHEC)
- Enterotoxigenic *E. coli* (ETEC)
- Enteropathogenic *E. coli* (EPEC)
- Enteroaggregative *E. coli* (EAggEC)
- Enteroinvasive *E. coli* (EIEC)
- Diffusely adherent *E. coli* (DAEC).

EHEC causes bloody diarrhoea, haemorrhagic colitis, haemolytic uraemic syndrome and thrombic thrombocytopenia purpura. This group includes the verotoxigenic *E. coli* (VTEC, also known as Shiga-toxin-producing *E. coli* or STEC) serotypes O157, O26 and O111.

Sources:

- EHEC: cattle (milk and meat products)
- Non-EHEC: humans.

Control measures:

- Effective sewage and water treatment
- Prevention of cross-contamination from raw foods and contaminated water
- Heat treatment: cooking, pasteurisation
- Good personal hygiene.

EHEC were first described in 1977 and recognised as a disease of animals and humans in 1982. The EHEC belong to many serogroups. The serotype O157:H7 is the most important in the UK and USA ('O157' and 'H7' refer to the serotyping of the strain's O- and H-antigens, respectively) and can cause very severe forms of food poisoning, resulting in death. It has been postulated that *E. coli* O157:H7 evolved from EPEC and acquired the toxin genes from *Sh. dysenteria* via a bacteriophage, and that the newly emerging pathogen arrived in Europe from South America (Coghlan 1998). The reported incidence and serotype varies from country to country. In the UK the only EHEC serotype recognised is O157:H7, whereas in France it is serotype O111.

Children and the elderly are the most vulnerable members of the population and develop haemorrhagic colitis (HC) which may lead to haemolytic uraemic syndrome (HUS). Healthy adults suffer from thrombic thrombocytopenic purpura (TTP) where blood platelets surround internal organs, leading to damage of the kidneys and central nervous system. HC is a less severe form of *E. coli* O157:H7 infection than HUS. The first symptom of HC is the sudden onset of severe crampy abdominal pains. About 24 hours later non-bloody (watery) diarrhoea starts. Some victims have a fever of short duration. Vomiting occurs in about half of the patients during the period of non-bloody diarrhoea and/or other times in the illness. After 1 or 2 days, the diarrhoea becomes bloody and the patient experiences increased abdominal pain. This usually lasts between 4 and 10 days. In severe cases, faecal specimens are described as 'all blood and no stool'. In most patients, the bloody diarrhoea resolves with no long-term impairment. Unfortunately, 2–7% patients (up to 30% in certain outbreaks) will progress to HUS and subsequent complications. Acute renal failure is the leading cause of death in children, whereas thrombo-cytopenia is the leading cause of death in adults. The Shiga-like toxins are specific for the glycosphingolipid globotriaosylceramide (Gb3) which is present on renal endothelial cells. Because Gb3 is found in the glomeruli of infants under 2 years of age but not in the glomeruli of adults, the presence of Gb3 in the pediatric renal glomerulus may be a risk factor for development of HUS.

In HUS the patient suffers from bloody diarrhoea, haemolytic anaemia, kidney disorder and renal failure, and requires dialysis and blood trans-fusions. Central nervous system disease may develop which can lead to seizures, coma and death. The mortality rate is 3–17%. HUS is a leading cause of kidney failure in children, which often requires dialysis and may ultimately be fatal. Other systemic manifestations of illness due to *E. coli* O157:H7 include a central nervous system involvement, hypertension, myocarditis and other cardiovascular complications that may result in death or severe disability. In some cases, the illness is indicative of some forms of heart disease and has been responsible for strokes in small chil-dren. These complications are attributed to direct or indirect actions of verotoxins absorbed from the intestinal tract.

Cattle appear to be the main reservoir of *E. coli* O157:H7. Transmission to humans is principally through the consumption of contaminated foods, such as raw or undercooked meat products and raw milk (see Table 1.2). Freshly pressed apple juice or cider, yogurt, cheese, salad vegetables and cooked maize have also been implicated. Faecal contamination of water and other foods, as well as cross-contamination during food preparation, are also implicated as transmission routes. There is evidence of trans-mission of this pathogen through direct contact between people. *E. coli* O157:H7 can be shed in faeces for a median period of 21 days with a range of 5–124 days.

According to the US FDA, the infectious dose for *E. coli* O157:H7 is unknown (FDA Bad Bug Book; see Internet Directory); however, a com-pilation of outbreak data indicates that it may be as low as 10 organisms. The data show that it takes a very low number of micro-organisms to cause illness in young children, the elderly and immuno-compromised people.

Most reported outbreaks of EHEC infection have been caused by

O157:H7 strains. This suggests that this serotype is more virulent or more transmissible than other serotypes. However, other serotypes of EHEC have been implicated in outbreaks, and the incidence of disease due to non-O157:H7 serotypes seems to be rising. More than 50 of these serotypes have been associated with bloody diarrhoea or HUS in humans. The most common non-O157:H7 serotypes associated with human disease include: O26:H11, O103:H2, O111:NM and O113:H21. At least ten of these outbreaks due to non-O157:H7 organisms have been reported in Japan, Germany, Italy, Australia, the Czech Republic, and the United States. These outbreaks have involved 5–234 people, and for most of them the source of infection could not be determined. In many countries, such as Chile, Argentina, and Australia, non-O157:H7 serotypes have been found to be responsible for the majority of HUS cases. Non-bloody diarrhoea has also been associated with some of these non-O157:H7 serotypes.

4.5.1 E. coli O157:H7 *in ground beef*

A model of *E. coli* O157:H7 in ground beef was constructed by Cassin *et al.* (1998a) to assess the impact of different control strategies. The model described the pathogen population from carcass processing through to consumer cooking and consumption. The paper also introduced the term 'process risk model' which combined quantitative risk assessment with scenario analysis and predictive microbiology. This approach was also used in the *C. jejuni* microbiological risk assessment (Section 4.3.1).

The study used two mathematical models:

(1) Model (1) described the behaviour of the pathogen from the production of the food through processing, handling and consumption to predict human exposure.
(2) The exposure estimate from model (1) was then used in a dose-response model to estimate the human health risk associated with consuming food from the process.

The prevalence and concentration of the pathogen was determined for each stage of the food chain ('farm to fork'). The effect of each stage on the prevalence and concentration of *E. coli* O157 is shown in Fig. 4.16. The figure starts with the prevalence (range 0/1131 to 188/11881) and concentration (range < 2.0–$5.0 \log_{10}$ cfu g^{-1}) data for *E. coli* O157:H7 in faeces.

The effect of processing and grinding on the prevalence and numbers of *E. coli* O157:H7 was then determined (see Fig. 4.16). For example, cross-contamination could increase the prevalence threefold, whereas spray

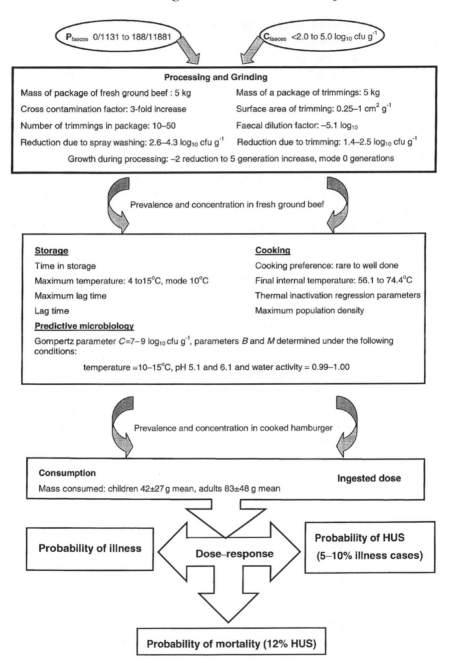

Fig. 4.16 Microbiological risk assessment for *E. coli* O157:H7 in hamburgers (adapted from Cassin *et al.* (1998a)).

washing could reduce the microbial numbers by 2.6–4.3 log orders. The predictive microbiology software Food Micromodel™ (Section 2.7.3) was used to simulate the growth of *E. coli* O157:H7 (modified Gompertz equation; Section 2.7.1). The amount of growth (parameter C) was assumed to be no greater than 7–$9\log_{10}$ cfu g^{-1}, the maximum rate of growth (parameter *B*) and length of time until maximum growth occurs (parameter *M*) were determined under the following conditions: temperature 10–15°C, pH 5.1 and 6.1, and water activity 0.99–1.00. The inactivation of *E. coli* O157 during cooking was estimated according to the range of consumer preferences: rare to well-done (54.4°C to 68.3°C internal temperature, respectively). Because of the range of estimates at each stage, Monte Carlo simulation was used (25 000 iterations for each simulation; Section 3.3.10) to generate a representative distribution of risk.

The ingested dose varied with age group, because adults consumed almost double the amount that children did (83 g compared with 42 g). The associated probability of exposure was calculated and a dose–response assessment constructed. The dose–response model was based on a modified beta-Poisson model for infection termed the beta-binomial model (see Section 3.3.6). It did not assume any threshold level and was based on α and β parameters similar to those of *Sh. dysenteriae* (Crockett *et al.* 1996). As can be seen in Fig. 4.17, there was considerable uncertainty in the probability of illness for a particular dose. From the dose–response analysis, the probability of illness, HUS and mortality were determined. The probability of HUS was 5–10% of the probability of illness cases, and the probability of mortality following HUS was assumed to be 12% of HUS cases (Fig. 4.16).

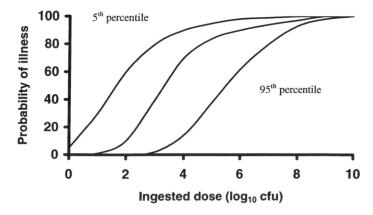

Fig. 4.17   Beta-Poisson dose–response model for *E. coli* O157:H7 (Cassin *et al.* 1998a).

The prevalence of packages containing *E. coli* O157 was estimated to be 2.9% and was comparable with surveillance data. The probability of illness per single hamburger meal ranged from 10^{-22} to 10^{-2} with a central tendency at 10^{-12}. Hence the majority of hamburger meals are predicted to have a very small risk to the consumer, but this is not negligible. The average probability of illness from a single meal was 5.1×10^{-5} for adults and 3.7×10^{-5} for children. Consequently, the model predicted a HUS probability of 3.7×10^{-6} and a mortality probability of 1.9×10^{-7} per meal for the very young. These values were deemed as representative of home-prepared hamburgers, and high for commercial production.

Spearman rank correlation coefficient (Fig. 4.18) was used to show that risk was most sensitive to the concentration of *E. coli* O157:H7 in faeces, and hence indicates the use of screening animals before slaughter as a control point.

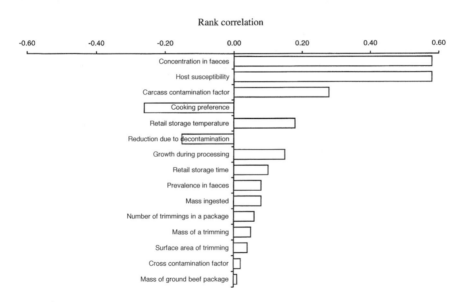

Fig. 4.18 Spearman rank correlation of risk factors for *E. coli* O157:H7 (Cassin *et al.* 1998a).

The model further enables changes in health risk associated with changes in control strategies to be predicted. The average probability of illness was predicted to be reduced by 80% by reducing microbial growth via reducing the storage temperature (Table 4.8). This was modelled on an average storage temperature of 8°C and a temperature abuse of no more than 13°C (compared with previous values of 10°C and 15°C, respectively). This risk reduction approach was predicted to be more effective

Table 4.8   Risk mitigation strategy, percentage reduction per meal illness from *E. coli* O157:H7 following assumed compliance (Cassin *et al.* 1998a).

Strategy	Control variable	Predicted reduction in illness (%)
(1) Storage temperature	Maximum storage temperature: 8°C mode, 13°C max	80
(2) Pre-slaughter screening	Concentration of *E. coli* O157:H7 in faeces reduced by 4 log orders	46
(3) Hamburger cooking Consumer information programme on cooking hamburgers	Cooking temperature, increase thorough cooking	16

than the reduction in the concentration of the *E. coli* O157:H7 in cattle faeces and even encouraging thorough cooking by the consumer. An online explanation of the FSIS model of *E. coli* in ground beef is available (see Internet Directory).

4.6   *Bacillus cereus*

Sources:

- Ubiquitous: soil, vegetation, meat, milk, water, rice, fish.

Control measures:

- Temperature control to prevent spore germination and outgrowth
- Storage temperature either $>60°C$ or $<10°C$, unless other parameters (e.g. pH, a_w) prevent bacterial growth
- Avoid storage of precooked foods.

B. cereus produces spores which can survive many cooking processes. The organism grows well in cooked food because of the lack of a competing microflora. *B. cereus* is ubiquitous in nature, being isolated from soil, vegetation, fresh water and animal hair. It is commonly found at low levels in food ($<10^2$ cfu g^{-1}) which is considered acceptable. Food poisoning outbreaks usually occur when the food has been subjected to time–temperature abuse which was sufficient for the low level of organisms to multiply to a significant (intoxication) level ($> 10^5$ cfu g^{-1}). Characteristics of temperature and water activity growth range and D values are given in Table 2.4.

There are two recognised types of *B. cereus* food poisoning: diarrhoeal and emetic. Both types of food poisoning are self-limiting and recovery usually occurs within 24 hours. *B. cereus* produces the diarrhoeal toxins during growth in the human small intestines, whereas the emetic toxins are preformed on the food and are heat-resistant (see Table 4.9).

Table 4.9   Characteristics of food poisoning caused by *B. cereus*.

	Emetic syndrome	Diarrhoeal syndrome
Infective dose	10^5–10^8 cells/g	10^5–10^7 total
Toxin produced	Preformed	Produced in the small intestine
Type of toxin	Cyclic peptide (1.2 kDa)	Three protein subunits (L$_1$, L$_2$, B; 37–105 kDa)
Toxin stability	Very stable (126°C, 90 min; pH 2–11)	Inactivated at 56°C, 30 minutes
Incubation period	0.5–6 hours	8–24 hours
Duration of illness	6–24 hours	12–24 hours
Symptoms	Abdominal pain, watery diarrhoea, nausea	Nausea, vomiting and malaise, water diarrhoea
Foods most frequently implicated	Fried and cooked rice, pasta, noodles and potatoes	Meat products, soups, vegetables, fish, puddings, sauces, and milk and dairy products

Adapted from Granum & Lund (1997)

Symptoms of *B. cereus* diarrhoeal food poisoning are:

- Watery diarrhoea
- Abdominal cramps and pain
- Nausea, rarely vomiting.

The symptoms of *B. cereus* diarrhoeal type food poisoning mimic those of *Cl. perfringens* food poisoning. The onset of watery diarrhoea, abdominal cramps, and pain occurs 8–24 hours after consumption of contaminated food. Nausea may accompany diarrhoea, but vomiting (emesis) rarely occurs. Symptoms persist for 24 hours in most instances, during which time the organism is excreted in large numbers.

Symptoms of *B. cereus* emetic food poisoning are:

- Nausea
- Vomiting
- Abdominal cramps and diarrhoea may occur.

The emetic type of food poisoning is characterised by nausea and vomiting 0.5–6 hours after consumption of contaminated foods. Occasionally, abdominal cramps and/or diarrhoea may also occur. Duration of symptoms is generally less than 24 hours. The symptoms of this type of food poisoning parallel those caused by *St. aureus* food-borne intoxication. Some strains of *B. subtilis* and *B. licheniformis* have been isolated from lamb and chicken incriminated in food poisoning episodes. These organisms demonstrate the production of a highly heat-stable toxin which may be similar to the vomiting type toxin produced by *B. cereus*.

A wide variety of foods, including meats, milk, vegetables and fish, have been associated with the diarrhoeal type food poisoning (Table 1.2). The vomiting-type outbreaks have generally been associated with rice products. However, other starchy foods such as potato, pasta and cheese products have also been implicated. Food mixtures, such as sauces, puddings, soups, casseroles, pastries, and salads, have frequently been incriminated in food poisoning outbreaks. The presence of large numbers of *B. cereus* (more than 10^6 cells g^{-1}) in a food is indicative of active growth and proliferation of the organism, and is consistent with a potential hazard to health.

Because the organism is ubiquitous in the environment, low numbers commonly occur in food. Therefore the main control mechanism is to prevent spore germination and multiplication in cooked, RTE foods. Storage of foods below 10°C will inhibit *B. cereus* growth.

4.6.1   B. cereus *risk assessment*

Notermans *et al.* (1997) and Notermans and Batt (1998) have described a risk assessment approach for *B. cereus*. The added difficulty in hazard characterisation and dose–response assessment of this organism is that it can cause two different illnesses according to the toxin(s) produced. Epidemiological studies by Kramer and Gilbert (1989) showed that the number of *B. cereus* organisms in food causing diarrhoeal and emetic food poisoning varied from 1.2×10^3 to 10^8 and 1.0×10^3 to 5.0×10^{10} cfu g^{-1}, respectively. The median for both was approximately 1×10^7 cfu g^{-1}. Human volunteer studies used pasteurised milk naturally contaminated with *B. cereus* (Langeveld *et al.* 1996; Fig. 4.19). Symptoms were observed with ingested doses of more than 10^8 cells. The dose–

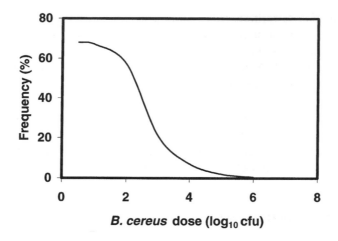

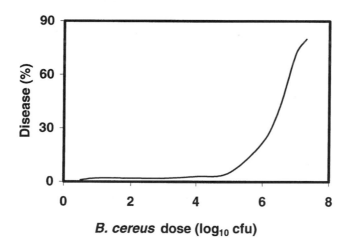

Fig. 4.19   *Bacillus cereus:* (a) Frequency in pasteurised milk; (b) dose–response curve.

response curve as given by Notermans and Mead (1996) is reproduced (generalised) in Fig. 4.19. There is probably considerable variation within *B. cereus* strains with regard to toxin production, and the amount of toxin produced may be food-related.

Because consumption of 10^4 *B. cereus* cells does not appear to be harmful (good manufacturing practice) and RTE levels are approximately 10^3 cfu g^{-1} (Anon. 1996), risk assessment needs to be applied with foods with levels greater than this value. This contrasts with the zero tolerance approach of *S.* Enteritidis risk assessment. Surveillance data show that a number of RTE products such as boiled rice, cream, herbs and spices, and even pasteurised milk, may have levels of *B. cereus* greater than $10^4 g^{-1}$ (te Giffel *et al.* 1997). Notermans *et al.* (1997) showed that 11% of milk consumed in The Netherlands contained more than 10 000 *B. cereus* cells per millilitre and that about approximately 10^9-10^{10} portions of pasteurised milk were consumed annually. Although to date there is insufficient distribution frequency data for a detailed quantitative risk assessment for *B. cereus*, Zwietering *et al.* (1996) have constructed a predictive growth model for the organism.

4.7 *Vibrio parahaemolyticus*

V. parahaemolyticus is recognised as a major cause of food-borne gastroenteritis in Japan. This is because the organism is associated with the consumption of seafoods which are a significant part of the average diet in Japan.

Typical symptoms of *V. parahaemolyticus* food poisoning are as follows:

• Diarrhoea
• Abdominal cramps
• Nausea
• Vomiting
• Headaches
• Fever and chills.

The incubation period is 4–96 hours after the ingestion of the organism, with a mean of 15 hours. The illness is usually mild or moderate, although some cases may require hospitalisation. On average, the illness lasts about 3 days. The infective dose is possibly greater than 1 million organisms. Virulence is due to the production of a thermostable direct haemolysin (TDH), the ability to invade enterocytes and possibly the production of an enterotoxin (possibly Shiga-like). The production of the *tdh* gene is the only reliable trait that currently distinguishes between pathogenic and non-pathogenic strains.

There are usually fewer than 10^3 cfu g^{-1} of *V. parahaemolyticus* on fish and shellfish, except in warm waters where the count may increase to 10^6 cfu g^{-1}. Infections with this organism have been associated with the consumption of raw, improperly cooked, or cooked, recontaminated fish and shellfish. A correlation exists between the probability of infection and warmer months of the year. Improper refrigeration of seafoods contaminated with this organism will allow its proliferation, which increases the possibility of infection. The organism is very sensitive to heat and outbreaks are thus frequently due to improper handling procedures and temperature abuse. Control of the organism is therefore through prevention of the organism's multiplication after harvesting by chilling (to below 5°C) and cooking to an internal temperature of more than 65°C. Isolation of any *Vibrio* species from cooked food indicates poor hygienic practice, because the organism is rapidly killed by heat.

4.7.1   Public health impact of V. parahaemolyticus in raw molluscan shellfish

The FDA (2000b) have released a draft risk assessment of *V. parahaemolyticus* in raw molluscan shellfish. The document (118 pages in length) can be downloaded (see Internet Directory). The report is sectioned according to the four risk assessment steps (hazard identification, exposure assessment, hazard characterisation and risk characterisation), and risk mitigation factors are also presented. The report states the various assumptions which have been made in order to elicit useful feedback. Dose–response analysis was included in both hazard characterisation and exposure assessment (post-harvest module).

An outline of the risk assessment is given in Fig. 4.20. The FDA risk assessment was in response to four outbreaks in 1997–1998 which totalled over 700 cases. The process of exposure assessment is summarised in Fig. 4.21. Information and data were gathered for three modules: harvest, post-harvest and public heath. The harvest module simulated the variation in total and pathogenic *V. parahaemolyticus* densities as a function of environmental conditions and showed that salinity was not an important parameter. Subsequently, the growth of the pathogen was modelled using only water temperature. The post-harvest model simulated the effect of current handling practices on the pathogen prevalence and concentration at time of consumption. The public health module estimated the distribution of the probable number of illnesses within one of five regions studied and season. The dose–response relationship was modelled using beta-Poisson, Gompertz and Probit relations (Section 3.3.6). The output from the three models was used to determine the risk of illness. For the Gulf Coast region the average number of illnesses predicted was 25, 1200, 3000 and 400 in the winter, spring, summer and autumn (fall), respectively. Average nationwide risk of illness was 4750 cases with a range of 1000–16 000.

The risk assessment model enabled the current FDA criterion of less than 10^4 *V. parahaemolyticus* per gram of shellfish to be evaluated. It was

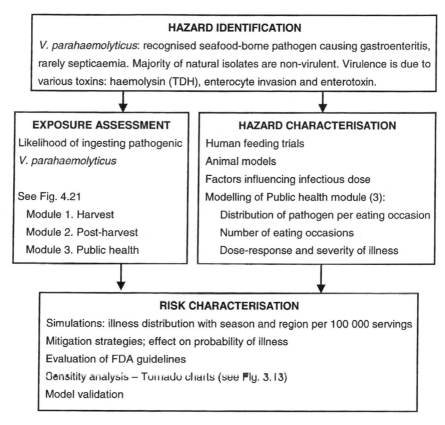

HAZARD IDENTIFICATION

V. parahaemolyticus: recognised seafood-borne pathogen causing gastroenteritis, rarely septicaemia. Majority of natural isolates are non-virulent. Virulence is due to various toxins: haemolysin (TDH), enterocyte invasion and enterotoxin.

EXPOSURE ASSESSMENT

Likelihood of ingesting pathogenic

V. parahaemolyticus

See Fig. 4.21

   Module 1. Harvest

   Module 2. Post-harvest

   Module 3. Public health

HAZARD CHARACTERISATION

Human feeding trials

Animal models

Factors influencing infectious dose

Modelling of Public health module (3):

   Distribution of pathogen per eating occasion

   Number of eating occasions

   Dose-response and severity of illness

RISK CHARACTERISATION

Simulations: illness distribution with season and region per 100 000 servings

Mitigation strategies; effect on probability of illness

Evaluation of FDA guidelines

Sensitity analysis – Tornado charts (see Fig. 3.13)

Model validation

Fig. 4.20 Risk assessment for *Vibrio parahaemolyticus* in raw molluscan shellfish (FDA 2000b).

estimated that excluding all oysters contaminated with more than 10^4 *V. parahaemolyticus* cells per gram would reduce associated illness by 15% and cause a 5% loss of harvest.

Risk mitigating strategies were (see Fig. 3.13):

- Cooling oysters immediately after harvest and keeping them refrigerated, during which time the viability of *V. parahaemolyticus* slowly decreased.
- Mild heat treatment (5 min, 50°C) giving a decrease in viability greater than 4.5 log and almost eliminating the likelihood of illness.
- Quick-freezing and frozen storage giving a 1–2 log decrease in viability and hence reducing the likelihood of illness.

The preliminary model can be downloaded (see Internet Directory).

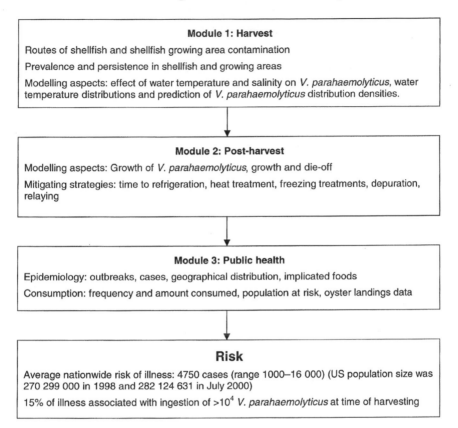

Module 1: Harvest

Routes of shellfish and shellfish growing area contamination

Prevalence and persistence in shellfish and growing areas

Modelling aspects: effect of water temperature and salinity on *V. parahaemolyticus*, water temperature distributions and prediction of *V. parahaemolyticus* distribution densities.

Module 2: Post-harvest

Modelling aspects: Growth of *V. parahaemolyticus*, growth and die-off

Mitigating strategies: time to refrigeration, heat treatment, freezing treatments, depuration, relaying

Module 3: Public health

Epidemiology: outbreaks, cases, geographical distribution, implicated foods

Consumption: frequency and amount consumed, population at risk, oyster landings data

Risk

Average nationwide risk of illness: 4750 cases (range 1000–16 000) (US population size was 270 299 000 in 1998 and 282 124 631 in July 2000)

15% of illness associated with ingestion of >10^4 *V. parahaemolyticus* at time of harvesting

Fig. 4.21 Exposure assessment of *Vibrio parahaemolyticus* in raw molluscan shellfish (FDA 2000b).

4.8   Mycotoxins

There are no detailed microbiological risk assessment studies of mycotoxins to present. However, due to their importance (severity) as food contaminants, a brief overview is included here. It should be noted that pertinent risk assessments are currently being undertaken (Moss 1999; Mantle 1999; Norman 1999; Scudamore 1999).

Mycotoxins are the toxic products of certain fungi which, in some circumstances, develop on or in foodstuffs of plant or animal origin (see Table 4.10). They are ubiquitous and widespread at all levels of the food chain (Webley *et al.* 1997; Li *et al.* 2000). Hundreds of mycotoxins have been identified and are produced by some 200 varieties of fungi of the genera *Aspergillus*, *Fusarium* and *Penicillium*. These fungi are ubiquitous and are part of the normal flora of plants.

Table 4.10 Toxicity of mycotoxins (adapted from Adams & Motarjemi (1999).

Mycotoxin	Food	Fungus species	Biological effect	LD_{50} (mg kg^{-1})
Aflatoxins	Maize, groundnuts, milk	*Asp. flavus, Asp. parasiticus*	Hepatotoxic, carcinogenic	0.5 (dog), 9.0 (mouse)
Cyclopiazonic acid	Cheese, maize, groundnuts	*Asp. flavus, Pen. aurantiogriseum*	Convulsions	36 (rat)
Fumonisin	Maize	*Fus. moniliforme*[a]	Equine encephalomalacia, pulmonary oedema in pigs	Unknown
Ochratoxin	Maize, cereals, coffee beans	*Pen. verruculosum, Asp. ochraceus*	Nephrotoxic	20–30 (rat)
Zearalenone	Maize, barley, wheat	*Fus. graminearum*[b]	Oestrogenic	Not acutely toxic

[a] *Gibberella fujikuroi.*
[b] *Gibberella zeale.*

Mycotoxins are secondary metabolites which have been responsible for major epidemics in man and animals. The aflatoxins (produced by *Aspergillus* spp.) range from single heterocyclic rings to six- or eight-membered rings. *Penicillium* produces a range of 27 mycotoxins such as patulin (an unsaturated lactone) and penitrem A (nine adjacent rings composed of 4–8 atoms). Ergotism, alimentary toxic aleukia, stachybotryotoxicosis and aflatoxicosis have killed thousands of humans and animals in the past century.

There are four types of toxicity:

- Acute, resulting in liver or kidney damage
- Chronic, resulting in liver cancer
- Mutagenic, causing DNA damage
- Teratogenic, causing cancer in the unborn child.

Epidemiological studies have shown a correlation between the high incidence of liver cancer in some African and South-East Asian countries (12–13 cases per 100 000 annually) and exposure to aflatoxins. Ochratoxin A has also been associated with Balkan endemic nephropathy, a fatal kidney disease prevalent in several Balkan countries.

4.8.1 Aflatoxins

The aflatoxins have been studied in more detail than other mycotoxins; they are a group of structurally related toxic compounds produced by certain strains of the fungi *Asp. flavus* and *Asp. parasiticus* under favourable conditions of temperature and humidity. These fungi grow on certain foods and feeds, where they produce aflatoxins. The most pronounced contamination has been encountered in tree nuts, peanuts and other oilseeds, including corn and cottonseed. The major aflatoxins of concern are designated B_1, B_2, G_1 and G_2 by the blue (B) or green (G) fluorescence given when viewed under a UV lamp. These toxins are usually found together in various foods and feeds in different proportions. However, aflatoxin B_1 is usually predominant and is the most toxic. Whether exposure is predominantly to aflatoxin B_1 or to mixed B_1 and G_1 depends on the geographical distribution of the *Aspergillus* strains. *Asp. flavus*, which produces aflatoxins B_1 and B_2, occurs world-wide; *Asp. parasiticus*, which produces aflatoxins B_1, B_2, G_1 and G_2, occurs principally in the Americas and in Africa. Exposure occurs primarily through dietary intake of maize and groundnuts. Aflatoxin M is a major metabolic product of aflatoxin B_1 in animals and is usually excreted in the milk and urine of dairy cattle and other mammalian species that have consumed aflatoxin-contaminated food or feed. Lifetime exposure to aflatoxins in some parts of the world, commencing *in utero*, has been confirmed by biomonitoring.

Aflatoxins produce acute necrosis, cirrhosis, and carcinoma of the liver in a number of animal species; no animal species is resistant to the acute toxic effects of aflatoxins (Table 4.10). Hence it is sensible to assume that humans may be similarly affected. A wide variation in LD_{50} values has been obtained in animal species tested with single doses of aflatoxins. For most species, the LD_{50} value ranges from 0.5 to 10 mg kg^{-1} bodyweight. Animal species respond differently in their susceptibility to the chronic and acute toxicity of aflatoxins. The toxicity can be influenced by environmental factors, exposure level and duration of exposure, age, health and nutritional status of diet. Aflatoxin B_1 is a very potent carcinogen in many species, including non-human primates, birds, fish and rodents. In each species, the liver is the primary target organ of acute injury. Metabolism plays a major role in determining the toxicity of aflatoxin B_1. This aflatoxin requires metabolic activation to exert its carcinogenic effect and these effects can be modified by induction or inhibition of the mixed function oxidase system.

In well-developed countries, aflatoxin contamination rarely occurs in foods at levels that cause acute aflatoxicosis in humans. In view of this, studies on human toxicity from ingestion of aflatoxins have focused on their carcinogenic potential. The relative susceptibility of humans to aflatoxins is not known, even though epidemiological studies in Africa and South-East Asia, where there is a high incidence of hepatoma, have revealed an association between cancer incidence, hepatitis B infection and the aflatoxin content of the diet.

The discovery of aflatoxins in the 1960s led to extensive surveying of koji moulds for mycotoxin production. Although under laboratory conditions mycotoxins can be produced by *Asp. oryzae*, *Asp. sojae* and *Asp. tamari*, no aflatoxins have been demonstrated in commercial production strains. The moulds used in cheese manufacture have also been tested for toxin production. *Penicillium roquefortii* produces trace amounts of patulin and roquefortine C, Whereas *P. camembertii* produces low levels of cyclopiazonic acid. These toxins are only produced under laboratory induced stress conditions, and it is reported that levels in cheese are 'extremely low' (Rowan *et al.* 1998).

Bowers *et al.* (1993) studied the possible link between aflatoxin exposure, hepatitis B prevalence and primary liver cancer in China. This included the data of Yeh *et al.* (1989) for almost 8000 men and aimed to construct a model for extrapolation for the USA. The best estimates of lifetime cancer potency of aflatoxin in the USA were 9 mg kg^{-1} day^{-1} and 230 mg kg^{-1} day^{-1} for hepatitis B-negative and -positive populations, respectively. This is lower than the $75\,mg\,kg^{-1}\,day^{-1}$ estimate for Africa and South-East Asia due to the greater prevalence of hepatitis B.

4.8.2   Ochratoxins

Ochratoxins are produced by *A. ochraceus*, *Penicillium verrucosum* and *P. viridicatum*. Ochratoxin A is the most potent of these toxins. The main dietary sources are cereals, but significant levels of contamination may be found in grape juice and red wine, coffee, cocoa, nuts, spices and dried fruits. Contamination may also carry over into pork and pig blood products and into beer. Ochratoxin is potentially nephrotoxic and carcinogenic, the potency varying markedly between species and sexes. It is also tetratogenic and immunotoxic. The JECFA have been asked by the CAC to perform a risk assessment on the levels of 5 and $20\,\mu g\,kg^{-1}$ of ochratoxin A in cereals and cereal products.

4.8.3   Other mycotoxins

Fumonisins are a group of *Fusarium* mycotoxins occuring world-wide in maize and maize-based products. Their causal role in several animal diseases has been established. Available epidemiological evidence has suggested a link between dietary fumonisin exposure and human oesophageal cancer in some locations with high disease rates. Fumonisins are mostly stable during food processing (DeNijs *et al.* 1997; Turner *et al.* 1999; Shier 2000).

Zearalenone is a fungal metabolite produced mainly by *F. graminearium* and *F. culmorum* which are known to colonise maize, barley, wheat, oats and sorghum. These compounds can cause hyperoestrogenism and severe reproductive and infertility problems in animals, especially in swine, but their impact on public health is hard to evaluate.

Trichothecenes are produced by many species of the genus *Fusarium*. They occur world-wide and infect many different plants, notably the cereal grains, especially wheat, barley and maize. There are over 40 different trichothecenes but the most well known are deoxynivalenol and nivalenol. In humans they cause vomiting, headache, fever and nausea.

Deoxynivalenol has been held responsible for large-scale human poisonings in China and India. Chronic exposures to deoxynivalenol, zearalenone and nivalenol occur in several parts of the world; humans are rarely exposed to fusarenone X. In episodes of food poisoning in humans caused by deoxynivalenol, severe gastrointestinal involvement was the primary symptom.

4.9 Rotaviruses

Rotaviruses are classified with the *Reoviridae* family. Six serological groups have been identified, three of which (groups A, B and C) infect humans (Desselberger 1996, 1998). Rotavirus gastroenteritis is a self-limiting, mild to severe disease characterised by vomiting, watery diarrhoea and low-grade fever (Hart & Cunliffe 1999; Ciarlet & Estes 2001). The infective dose is presumed to be 10–100 infectious viral particles (Shaw 2000; Lundgren & Svensson 2001). Because a person with rotavirus diarrhoea often excretes large numbers of virus (10^8–10^{10} infectious particles/ml of faeces) infectious doses can be readily acquired through contaminated hands, objects or utensils (Ponka *et al.* 1999). Asymptomatic rotavirus excretion has been well documented and may play a role in perpetuating endemic disease.

Rotaviruses are transmitted by the faecal–oral route. Person-to-person spread through contaminated hands is probably the most important means by which rotaviruses are transmitted in close communities such as paediatric and geriatric wards, daycare centres and family homes. Infected food handlers may contaminate foods that require handling and no further cooking, such as salads, fruits and hors d'oeuvres (Richards 2001). Rotaviruses are quite stable in the environment and have been found in estuary samples at levels as high as 1–5 infectious particles per gallon. Sanitary measures adequate for bacteria and parasites seem to be ineffective in endemic control of rotavirus, as a similar incidence of rotavirus infection is observed in countries with both high and low health standards (Mead *et al.* 1999; Fleet *et al.* 2000; Inouye *et al.* 2000; Sethi *et al.* 2001).

Group A rotavirus is endemic world-wide. It is the leading cause of severe diarrhoea among infants and children, and accounts for about half of the cases requiring hospitalisation. Over 3 million cases of rotavirus gastroenteritis occur annually in the USA. In temperate areas, it occurs primarily in the winter, but in the tropics it occurs throughout the year. The number attributable to food contamination is unknown. Group B rotavirus, also called adult diarrhoea rotavirus or ADRV, has caused major epidemics of severe diarrhoea affecting thousands of persons of all ages in China. Group C rotavirus has been associated with rare and sporadic cases of diarrhoea in children in many countries. However, the first outbreaks were reported from Japan and England. The incubation period ranges from 1 to 3 days. Symptoms often start with vomiting followed by 4–8 days of diarrhoea. Temporary lactose intolerance may occur. Recovery is usually complete. However, severe diarrhoea without fluid and electrolyte replacement may result in severe dehydration and death. Childhood mortality caused by rotavirus is relatively low in the USA, with an estimated 100 cases per year, but reaches almost 1 million cases per year world-wide. Association with other enteric pathogens may play a role in the severity of the disease.

4.9.1   Viral contamination of shellfish and coastal waters

Rose and Sobsey (1993) completed a four stage (*Codex* format) quantitative risk assessment of viral contamination of shellfish and coastal waters. The hazard identification stage summarised the viral hazards associated with contaminated shellfish: polivirus, echovirus and rotavirus. The dose–response assessment (part of hazard characterisation) used previous human feeding trial data (see Table 3.7) for echovirus 12 (low infectivity) and rotavirus (high infectivity). Exposure assessment was determined by multiplying the number of viral particles per gram of shellfish by the amount ingested per year. The average number of viral particles per gram of shellfish (range 0.2–31 pfu per 100 g) was calculated and the arithmetic average obtained for each sampling site. The amount of shellfish consumed per person per year in the USA was 250 g of clams, 74 g of oysters and 53 g of other shellfish. It was assumed that individuals consumed equal amounts of shellfish through the year. An average serving was determined to be 60–240 g. Hence exposures to a single serving ranged from 0.11–18.6 pfu (average 6 pfu) for a 60 g serving to 0.43–74.4 pfu for a 240 g serving. The risk characterisation determined risk estimates using the beta-Poisson probability models of echovirus 12 and rotavirus, virus contamination level and the two levels of consumption to represent low and high exposures. Illness and death rates were compared over a range of exposure doses and were multiplied by the probability of infection to determine the risk of illness and death. Using the echovirus 12 probability model for the consumption of 60 g of raw shellfish, the individual risks ranged from 2.2×10^{-4} to 3.5×10^{-2}, the risk being four times greater for 240 g servings. Using average virus levels, this equates to approximately a 1 in 100 chance of being infected following ingestion of raw shellfish from approved waters. However, using the rotavirus model the risk increased to 3.1×10^{-1} due to its greater infectivity with an average virus exposure of 6 pfu per 60 g of shellfish.

This early risk assessment understandably does not use the distribution range of viruses in a Monte Carlo simulation. It is also plausible that the use of improved detection methods would increase the virus incidence and level in shellfish compared to that used in this study.

5

FUTURE DEVELOPMENTS IN MICROBIOLOGICAL RISK ASSESSMENT

5.1 Introduction

The area of microbiological risk assessment has developed very rapidly from its early applications in the 1990s. It is therefore difficult to publish a predictive list which will not age too rapidly. Hence this chapter concentrates on topics requiring further investigation.

If microbiological risk assessment proves to be effective in facilitating the world-wide distribution of 'safe' food products, then it could subsequently enhance world trade in food. Additionally, microbiological risk assessment could lead to improved HACCP schedules due to the establishment of CCPs which reduce the microbial hazard to a scientifically justifiable 'acceptable' level (Section 2.3.4).

5.2 International methodology and guidelines

To facilitate the international adoption of microbiological risk assessment it will be necessary to:

(1) Produce an internationally agreed methodology
(2) Produce guidelines on hazard characterisation, exposure assessment and risk characterisation to give detailed guidance regarding information and data requirements and how to evaluate such information
(3) Agree on dose–response models.

In order to propagate risk assessment experience, collaborative projects involving experienced and inexperienced countries in a joint risk assessment are required. A means of presenting and disseminating the results of preliminary and completed microbiological risk assessments would support these initiatives.

More national microbiological risk assessments are required, in contrast to the current ones which largely rely on international data (see Denmark, *C. jejuni* in Section 4.3.3).

5.3   Data

A database of international risk assessment information should be established. This database should include data from developing countries. It is expected that time and temperatures of storage, preparation and cooking practices will differ between countries. Countries from all regions of the world need to report local practices of storage and handling of food in the home by consumers and in food-service establishments, including storage temperatures and times. Consumption details, such as size of portion and frequency of consumption, are also needed. The system of data input from industry should be such that there would be no punitive action by government nor use by competing companies.

Adoption of microbiological risk assessment is likely to increase the need for data on the presence and concentration of micro-organisms in food products to validate the risk assessment model(s). As already shown risk assessments of *Salmonella* (and *L. monocytogenes*) are restricted as a result of the standard practice of the detection criterion being the presence or absence of one cell in 25 g of food, rather than enumeration. Quantitative data on the concentration of pathogens, such as *Salmonella* on poultry carcasses, are required from all regions of the world. Cross-contamination determinations in the processing and preparation stages are needed for many microbial pathogens in the exposure assessment step.

Surveillance systems need to be established to support epidemiological and outbreak investigations. These data are needed for dose–response assessment models with populations of different susceptibilities. Sentinel studies are required to estimate more accurately the number of people with food-borne illness each year.

Predictive microbiology has to date largely focused on microbial growth (temperature abuse, etc.) and death (pasteurisation, cooking). There is a need to develop further predictive microbiology as a core facility and enable different countries to apply the models to their own farm-to-fork processing conditions. Predictive microbiology needs to be extended to account for:

- Processes with fluctuating time and temperature regimes
- Bacterial survival at chill and frozen temperatures
- Bacterial adaptation during processing; see Section 2.6.

In future, the dose–response models should represent:

(1) The probability of infection given exposure
(2) The probability of illness given infection
(3) The probability of sequelae and/or mortality given illness.

However, the dose–response curves for non-Typhi *Salmonella* (Fig. 4.4) demonstrate the diversity of mathematical models which currently can be applied to the same set of data. Hence an internationally agreed methodology for dose–response analysis is needed in the future (albeit that different models are used for different microbial hazards), so that one could predict the likely number of people becoming infected or ill from the number of bacteria (etc.) in food (food safety objective/microbiological criteria). The other important part of this will be to find new ways of obtaining relevant human data to guide our models, a process in which improved outbreak investigations will lead the way.

Food safety objectives and microbiological criteria need to be set which incorporate the risk associated with operating characteristic curves; see Section 2.8.2.

5.4 Training courses and use of resources

Risk assessment requires a multidisciplinary approach involving, for example, risk assessors, statisticians, microbiologists and epidemiologists. However, such skilled and experienced personnel are not distributed world-wide. On a global scale the creation of networks via the Internet would be valuable for the transfer of information, technical advice and collaborative studies. The JEMRA has already been established by the FAO/WHO for the provision of expert advice on microbiological risk assessment.

There is an increasing number of training courses or explanations of risk models becoming available from the Internet (see FAO/WHO in the Internet Directory under the heading 'Online training courses and model explanations' for all topics listed below).

• FAO/WHO initiative on microbial risk assessment (IAFP 88th Annual Meeting, August 2001)
• Overview of FAO/WHO initiative
• Exposure assessment of *Salmonella* spp. in broilers
• Exposure assessment of *Salmonella* Enteritidis in eggs
• Hazard and risk characterisation of *Salmonella*

- Exposure assessment of *L. monocytogenes* in ready-to-eat meat and fish
- Exposure assessment of *L. monocytogenes* in dairy products
- Hazard and risk characterisation of *L. monocytogenes*.

See WHO in the Internet Directory under the heading 'Online training courses and model explanation' for the following topics:

- Global perspective of risk assessment
- Quantitative risk assessment of *C. jejuni* in chicken
- Risk assessment advice
- Monte Carlo simulations
- *S.* Enteritidis in egg and egg products
- Quantitative risk modeling
- FSIS *E. coli* O157:H7 risk assessment in ground beef
- Poultry FARM.

Table 5.1   Proposed microbiological risk assessments (32nd CCFH meeting 1999 (CAC 2001)).

Hazard	Commodity
S. Enteritidis	Eggs
Salmonella spp.	Poultry, red meat, bean sprouts, fish
C. jejuni	Poultry
Enterohaemorrhagic *E. coli*	Beef, bean sprouts
L. monocytogenes	Soft cheese, ready-to-eat products, smoked fish, minimally processed vegetables (i.e. salads and precooked frozen vegetables)
V. parahaemolyticus	Shellfish
Vibrio spp.	Seafood
Cyclospora	Fresh produce
Shigella spp.	Vegetables
St. aureus	
B. cereus	Infant formula milk
Clostridium perfringens	
Cryptosporidium	
Small round structured viruses	
Antimicrobial resistance	

These need to be further developed with working examples of micro-biological risk assessment. It should be recognised, however, that the Internet is not available world-wide and alternative effective communication methods must be made available to all who need them, especially in developing countries.

Future microbiological risk assessments were identified by the 32nd

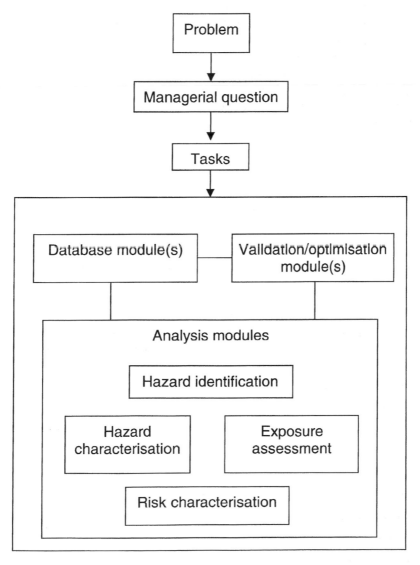

Fig. 5.1 Microbiological risk assessment support (MRAS) system (adapted from Jouve (2001)).

session of the CCFH (1999). Many, but not all, of these were pathogen-commodity combinations (Table 5.1). Because of their public health significance priority was given to the following:

(1)   *Salmonella* in eggs, poultry and pork meat
(2)   *L. monocytogenes* in ready-to-eat food
(3)   *C. jejuni* in poultry
(4)   Enterohaemorrhagic *E. coli* in brussels sprouts and ground beef
(5)   *V. parahaemolyticus* in shellfish.

Some of these risk assessments have been completed, whilst others were initiated in 2001 (Section 3.2) or will be started in the near future.

5.5   Microbiological risk assessment support system

Jouve (2001) proposed the microbiological risk assessment support system (MRAS) to help national analysts and decision makers (Fig. 5.1). It would use the experience gained in a few countries to assist others. Microbiological risk assessment would be composed of three groups of modules:

(1)   Analysis modules
(2)   Database module
(3)   Validation/optimisation modules.

The analysis modules would include the four activities of risk assessment, hazard identification, exposure assessment and risk characteristion. These require modelling and simulations for which various items of software are already available. The database module would cover guidance for data collection and incorporate pre-existing data. This would provide a template for conducting risk assessments and hence would be resource-efficient. The validation/optimisation module would provide tests for fitting models to data and attendant uncertainties.

GLOSSARY OF TERMS

These definitions are mainly those adopted on an interim basis at the 22nd session of the Codex Alimentarius Commission for microbiological, chemical, or physical agents and risk management and risk communication. The CAC adopted these definitions on an interim basis because they are subject to modification in the light of developments in the science of risk analysis and as a result of efforts to harmonise similar definitions across various disciplines.

Adverse effect: change in morphology, physiology, growth, development or lifespan of an organism which results in impairment of functional capacity or impairment of capacity to compensate for additional stress or increase in susceptibility to the harmful effects of other environmental influences. Decisions on whether or not any effect is adverse require expert judgement.

Assumption: an expert judgement made on the basis of incomplete information, which therefore has uncertainty associated with it.

***D* value (decimal reduction time):** 90% (= 1 log) loss of viability due to a lethal process such as heat (cooking), acidity or irradiation. See also *Z* value.

Dose–response assessment: the determination of the relationship between the magnitude of exposure (dose) to a chemical, biological or physical agent and the severity and/or frequency of associated adverse health effects (response).

Exposure assessment: the qualitative and/or quantitative evaluation of the likely intake of biological, chemical and physical agents via food as well as exposure from other sources if relevant.

Food: any substance, whether processed, semiprocessed or raw, which is intended for human consumption, including drinks, chewing gum and any substance which has been used in the manufacture, preparation or treatment of 'food', but excluding cosmetics, tobacco and substances used only as drugs.

Food safety objective: a government-defined target considered necessary to protect the health of consumers (this may apply to raw materials, a process or finished products).

Hazard: a biological, chemical or physical agent in, or condition of, food with the potential to cause an adverse health effect.

Hazard characterisation: the qualitative and/or quantitative evaluation of the nature of the adverse health effects associated with biological, chemical and physical agents which may be present in food. For the purpose of microbiological risk assessment the concerns relate to microorganisms and/or their toxins. For biological agents, a dose–response assessment should be performed if the data are obtainable.

Hazard identification: the identification of biological, chemical and physical agents capable of causing adverse health effects and which may be present in a particular food or group of foods.

Qualitative risk assessment: a risk assessment based on data which, whilst forming an inadequate basis for numerical risk estimate, nonetheless, when conditioned by prior expert knowledge and identification of attendant uncertainties, permits risk ranking (comparison) or separation into descriptive categories of risk.

Quantitative risk assessment: a risk assessment that provides numerical expression of risk and indication of the attendant uncertainties.

Microbiological risk: a risk that is related to the presence of a microbiological hazard (such as bacteria, viruses, yeast, moulds and algae, parasitic protozoa and helminths). This includes the chemical hazards they may produce (toxins and metabolites). (From proposed draft principles and guidelines for the conduct of microbiological risk assessment – CCFH 2000); (see Internet Directory.)

Risk: a function of the probability of an adverse health effect and the severity of that effect, consequential to a hazard(s) in food.

Risk analysis: a process consisting of three components: risk assessment, risk management and risk communication.

Risk assessment: a scientifically based process consisting of the following steps: (1) hazard identification, (2) hazard characterisation, (3) exposure assessment, and (4) risk characterisation. The definition includes quantitative risk assessment, which emphasises reliance on numerical expressions of risk, and also qualitative expressions of risk, as well as an indication of the attendant uncertainties.

Risk assessment policy: consists of documented guidelines for scientific judgement and policy choices to be applied at appropriate decision points during risk assessment.

Risk characterisation: the qualitative and/or quantitative estimation, including attendant uncertainties, of the probability of occurrence and severity of known or potential adverse health effects in a given population based on hazard identification, hazard characterisation and exposure assessment.

Risk communication: the interactive exchange of information and opinions throughout the risk analysis process concerning hazards and risks, risk-related factors and risk perceptions, among risk assessors, risk managers, consumers, industry, the academic community and other interested parties, including the explanation of risk assessment findings and the basis of risk management decisions.

Risk estimate: output of risk characterisation.

Risk management: the process, distinct from risk assessment, of weighing policy alternatives, in consultation with all relevant parties, considering risk assessment and other factors relevant for the health protection of consumers and for the promotion of fair trade practices, and, if needed, selecting appropriate prevention and control options.

Scenario set: a construct characterising the range of likely pathways affecting the safety of the food product. This may include consideration of processing, inspection, storage, distribution and consumer practices. Probability and severity values are applied to each scenario.

Sensitivity analysis: a method used to examine the behaviour of a model by measuring the variation in its outputs resulting from changes to its inputs.

Threshold: dose of a substance or exposure concentration below which a stated effect is not observed or expected to occur.

Transparent: characteristics of a process where the rationale, the logic of development, constraints, assumptions, value judgements, decisions, limitations and uncertainties of the expressed determination are fully and systematically stated, documented, and accessible for review.

Uncertainty: lack of sufficient or reliable data or knowledge.

Uncertainty analysis: a method used to estimate the uncertainty associated with model inputs, assumptions and structure and/or form.

Variability: distribution of values due to known variables such as biological variation, seasonal changes and amount of food eaten.

Z value: temperature increase required to increase the death rate tenfold, i.e. the temperature increase that reduces the D value tenfold.

REFERENCES

Adams, M. & Motarjemi, Y. (1999) *Basic Food Safety for Health Workers*. WHO/ SDE/PHE/FOS/99.1. World Health Organisation, Geneva.

Adams, M. R., Little, C. L. & Easter, M. C. (1991) Modelling the effect of pH, acidulant and temperature on the growth rate of *Yersinia enterocolitica. J. Appl. Bacteriol.* **71**, 65.

Allos, B. M. (1998) *Campylobacter jejuni* infection as a cause of the Guillain-Barré syndrome. *Emerg. Infect. Dis.* **12**, 173–184.

Altekruse, S. F., Tollefson, L. K. & Bögel, K. (1993) Control strategies for *Salmonella* Enteritidis in five countries. *Food Control* **4**, 10–16.

Anon. (1996) Microbiological guidelines for some ready-to-eat foods sampled at the point of sale: an expert opinion from the Public Health Laboratory Service (PHLS). *PHLS Microbiol. Digest* **13**, 41–43.

Anon. (1999a) Where have all the gastrointestinal infections gone? *EuroSurveill. Wk. Rep.* **3**, 990114.

Anon. (1999b) Opinion of the scientific committee on veterinary measures relating to public health on the evaluation of microbiological criteria for food products of animal origin for human consumption. See Internet site europa.eu.int/comm/food/fs/sc/scv/out26_en.pdf

Anon. (1999c) Egg safety, from production to consumption: an action plan to eliminate *Salmonella* Enteritidis illness due to eggs. President's Council on Food Safety, USA, 10 December 1999. See Internet site. www.foodsafety.gov/ ~fsg/ceggs.html

ANZFA (1999) *Food Safety Standards - Costs and Benefits*. Australia and New Zealand Food Authority. See Internet site www.anzfa.gov.au

Archer, D. L. & Young F. E. (1988) Contemporary issues: diseases with a food vector. *Clin. Microbiol. Rev.* **1**, 377–398.

Baker, D. A. (1995) Application of modelling in HACCP plan development. *Int. J. Food Microbiol.* **25**, 251–261.

Baik, H. S., Bearson, S., Dunbar, S. & Foster, J. W. (1996) The acid tolerance response of *Salmonella typhimurium* provides protection against organic acids. *Microbiology* **142**, 3195–3200.

Baker, D. A. & Genigeogis, C. (1990) Predicting the safe storage of fresh fish under modified atmospheres with respect to *Clostridium botulinum* toxigenesis by modeling length of the lag phase of growth. *J. Food Protect.* **53**, 131–140.

Baranyi, J. & Roberts, T. A. (1994) A dynamic approach to predicting bacterial growth in food. *Int. J. Food Microbiol.* **23**, 277–294.

Barker, G. C., Talbot, N. L. X. & Peck, M. W. (1999) Microbiological risk assessment for sous-vide foods. In: *Proceedings of Third European Symposium on Sous-Vide*, 25-26 March 1999, pp. 37-46. Alma Sous Vide Centre, Belgium.

Bauman, H. E. (1974) The HACCP concept and microbiological hazard categories. *Food Technol.* **28**, 30-34, 74.

Bemrah, N., Sana, M., Cassin, M. H., Griffiths, M. W. & Cerf, O. (1998) Quantitative risk assessment of human listeriosis from consumption of soft cheese made from raw milk. *Prevent. Vet. Med.* **37**, 129-145.

Bennett, J. V., Homberg, S. D., Rogers, M. F. & Solomon, S. L. (1987) Infectious and parasitic diseases. *Am. J. Prevent. Med.* **55**, 102-114.

Berends, B. R., van Knapen, F., Snijders, J. M. A., Mossel, D. A. A. (1997) Identification and quantification of risk factors regarding *Salmonella* spp. in pork carcasses. *Int. J. Food Microbiol.* **36**, 199-206.

Bettcher, D. W., Yach, D. & Guindon, G. E. (2000) Global trade and health: key linkages and future challenges. *Bull. WHO* **78**, 521-534.

Black, R. E., Levine, M. M., Clements, M. L., Highes, T. P. & Blaster, M. J. (1988) Experimental *Campylobacter jejuni* infection in humans. *J. Infect. Dis.* **157**, 472-479.

Booth, I. R., Pourkomailian, B., McLaggan, D. & Koo, S.-P. (1994) Mechanisms controlling compatible solute accumulation: a consideration of the genetics and physiology of bacterial osmoregulation. *J. Food Eng.* **22**, 381-397.

Borch, E. & Wallentin, C. (1993) Conductance measurement for data generation in predictive modelling. *J. Ind. Microbiol.* **12**, 286.

Bowers, J., Brown, B., Springer, J., Tollefson, L., Lorentzen, R. & Henry, S. (1993) Risk assessment for aflatoxin: an evaluation based on the multistage model. *Risk Anal.* **13**, 637-642.

Boyd, E. F., Wang, F. S., Whitham, T. S. & Selander, R. K. (1996) Molecular relationship of the salmonellae. *Appl. Env. Microbiol.* **62**, 804-808.

Broughall, J. & Brown, C. (1984) Hazard analysis applied to microbial growth in foods: development and application of three-dimensional models to predict bacterial growth. *J. Food Microbiol.* **1**, 13-22.

Broughall, J. W., Anslow, P. A. & Kilsby, D. C. (1983) Hazard analysis applied to microbial growth in foods: development of mathematical models to predict bacterial growth. *Food Microbiol.* **1**, 12-22.

Brown, M. H., Davies, K. W., Billoon, C. M. P., Adlair, C. & McClure, P. J. (1998) Quantitative microbiological risk assessment: principles applied to determining the comparative risk of salmonellosis from chicken products. *J. Food Protect.* **61**, 1446-1453.

Brown, P., Will, R. G., Bradley, R., Asher, D. & Detwiler, L. (2001) Bovine spongiform encephalopathy and variant Creutzfeld-Jakob disease: background, evolution and current concerns. *Emerg. Inf. Dis.* **7**, 6-16.

Bruce, M. E., Will, R. G., Ironside, J. W. *et al.* (1997) Transmission to mice indicates that 'new variant' CJD is caused by the BSE agent. *Nature* **389**, 498-501.

Brundtland, G. H. (2001) 24th Session of the Codex Alimentarius Commission, Geneva, 2 July 2001. See Internet site www.who.int/fsf.

Buchanan, R. L. (1995) The role of microbiological criteria and risk assessment in HACCP. *Food Microbiol.* **12**, 421-424.

Buchanan, R. L. (1997) National Advisory Committee on Microbiological Criteria for Foods. Principles of risk assessment for illnesses caused by foodborne biological agents. *J. Food Protect.* **60**, 1417-1419.

Buchanan, R. L., Damert, W. G., Whiting, R. C. & Van Schothorst, M. (1997) Use of epidemiologic and food survey data to estimate a purposefully conservative dose–response relationship for *Listeria monocytogenes* levels and incidence of listeriosis. *J. Food Protect.* **60**, 918–922.

Buchanan, R. L. & Edelson, S. G. (1999) Effect of pH-dependent, stationary phase acid resistance on the thermal tolerance of *Escherichia coli* O157:H7. *Food Microbiol.* **16**, 447–458.

Buchanan, R. L. & Lindqvist, R. (2000) Preliminary report. Hazard identification and hazard characterization of *Listeria monocytogenes* in ready-to-eat foods. See Internet site www.fao.org/WAICENT/FAOINFO/ECONOMIC/ESN/pager-isk/mra001.pdf

Buchanan, R. L., Smith, J. L. & Long, W. (2000) Microbial risk assessment: dose–response relations and risk characterization. *Int. J. Food Microbiol.* **58**, 159–172.

Buchanan, R. L. & Whiting, R. (1996) Risk assessment and predictive microbiology. *J. Food Protect.* **59**, Suppl., 31–36.

Buchanan, R. L. & Whiting, R. C. (1998). Risk assessment: a means for linking HACCP plans and public health. *J. Food Protect.* **61**, 1531–1534.

Bunning, V. K., Lindsay, J. A. & Archer, D. L. (1997) Chronic health effects of foodborne microbial disease. *World Health Stat. Quart.* **50**, 51–56.

Buzby, J. C. & Roberts, T. (1997a) Economic costs and trade implications of microbial foodborne illness. *World Health Stat. Quart.* **50**, 57–66.

Buzby, J. C. & Roberts, T. (1997b) Guillain–Barré syndrome increases foodborne disease costs. *Food Rev* **20**, 36–42.

CAC (1993) Codex Guidelines for the Application of the Hazard Analysis Critical Control Point (HACCP) System. Joint FAO/WHO Codex Committee on Food Hygiene. WHO/FNU/FOS/93.3 Annex II. Codex Alimentarius Commission, FAO, Rome.

CAC (1997a) *Hazard Analysis and Critical Control Point (HACCP) System Principles and Guidelines for its Application.* Annex II CAC/RCP-1 1969, Rev 3 M 99/13A Appendix II. Codex Alimentarius Commission, FAO, Rome.

CAC (1997b) Recommended International Code of Practice on the general principles of food hygiene. Annex II CAC/RCP-1 1969, Rev 3 M 99/13A Appendix II. Codex Alimentarius Commission, Rome.

CAC (1997c) Report of the Twenty-second Session, Codex Alimentarius Commission, Joint FAO/WHO Foods Standards Programme, Geneva, 23–28 June 1997. Alinorm 97/37. Codex Alimentarius Commission, FAO, Rome.

CAC (1998) *Draft Principles and Guidelines for the Conduct of Microbiological Risk Assessment.* Alinorm 99/13A Appendix II. Codex Alimentarius Commission, FAO, Rome.

CAC (1999) *Principles and Guidelines for the Conduct of Microbiological Risk Assessment.* CAC/GL-30. Codex Allimentarius Commission. FAO, Rome.

CAC (2001) Report of the 32nd meeting of the Codex Committee on Food Hygiene, 1999. See Internet site ftp://ftp.fao.org/codex/ALINORM01/A101_13e.pdf

Carlin, F., Girardin, H., Peck, M. W. *et al.* (2000) Research on factors allowing a risk assessment of spore-forming pathogenic bacteria in cooked chilled foods containing vegetables: a FAIR collaborative project. *Int. J. Food Microbiol.* **60**, 117–135.

Cassin, M. H., Lammerding, A. M., Todd, E. C. D., Ross, W. & McColl, R. S. (1998a) Quantitative risk assessment for *Escherichia coli* O157:H7 in ground beef hamburgers. *Int. J. Food Microbiol.* **41**, 21–44.

Cassin, M. H., Paoli, G. M. & Lammerding, A. M. (1998b) Simulation modeling for microbial risk assessment. *J. Food Protect.* **61**, 1560–1566.

CCFH (1998) Discussion paper on recommendations for the management of microbiological hazards for food in international trade. CX/FX/98/10. Codex Committee on Food Hygiene. FAO, Rome.

CCFH (1999) Management of *Listeria monocytogenes* in foods. Joint FAO/WHO Food Standards Programme, Codex Committee on Food Hygiene. 32nd Session. CX/FH 99/10 p. 27. FAO, Rome.

CCFH (2000) *Proposed Draft Principles and Guidelines for the Conduct of Microbiological Risk Management at Step 3.* Joint FAO/WHO Food Standards Programme, Codex Committee on Food Hygiene, July 2000. CX/FH 00/06. FAO, Rome.

CCFRA (2000) *An Introduction to the Practice of Microbiological Risk Assessment for Food Industry Applications.* Guideline 28. Campden & Chorleywood Food Research Association Group, Leatherhead, UK.

Christensen, B., Rosenquist, H., Sommer, H. & Nielsen N. (2001) Quantitative risk assessment of human illness associated with *Campylobacter jejuni* in broilers. Campylobacter, Helicobacter and related organisms (CHRO) 11th Workshop. Freiberg, Germany. See Internet site www.lst.min.dk/publikationer/publikationer/publikationer/campuk/cameng_ref.doc

Ciarlet, M. & Estes, M. K. (2001) Rotavirus and calicivirus infections of the gastrointestinal tract. *Curr. Opin. Gastroenterol.* **17**, 10–16.

Coghlan, A. (1998) Deadly *E. coli* strains may have come from South America. *New Scientist*, 10 January, p12.

Coleman, M. & Marks, H. (1998) Topics in dose–response modelling. *J. Food Protect* **61**, 1550–1559.

Collinge, J., Sidle, K. C. L., Meads, J., Ironside, J. & Hill, A. F. (1996) Molecular analysis of prion strain variation and the aetiology of 'new variant' CJD. *Nature* **383**, 685–690.

Corlett, D. A. (1998) *HACCP User's Manual.* Aspen, Gaithersburg, MD.

Crockett, C. S., Haas, C. N., Fazil, A., Rose, J. B. & Gerba, C. P. (1996) Prevalence of shigellosis in the US: consistency with dose–response information. *Int. J. Food Microbiol.* **30**, 87–99.

Crutchfield, S. R., Buzby, J. C., Roberts, T. & Ollinger, M. (1999) *FoodReview* **22**, 6–9.

DeNijs, M., van Egmond, H. P., Rombouts, F. M. & Notermans, S.H.W. (1997) Identification of hazardous *Fusarium* secondary metabolites occurring in raw food materials. *J. Food Safety* **17**, 161–191.

Desselberger, U. (1996) Genome rearrangements of rotaviruses. *Arch. Virol.* **11**, 37–51.

Desselberger, U. (1998) Viral gastroenteritis. *Curr. Opin. Infect. Dis.* **11**, 565–575.

Ebel, E. D., Kasuga, F., Schlosser, W. & Yamamoto, S. (2000) Exposure assessment of *Salmonella* Enteritidis in eggs. JEMRA. See Internet site www.fao.org/WAICENT/FAOINFO/ECONOMIC/ESN/pagerisk/mra004.pdf

Elliott, P. H. (1996) Predictive microbiology and HACCP. *J. Food Protect.* (Suppl), 48–53.

European Commission (1999) Opinion on principles for the development of risk assessment of microbiological hazards under the hygiene of foodstuffs Directive 93/43/EEC. See Internet site http://europa.eu.int/comm/food/fs/sc/scv/out26_en.pdf

Ewing, W. H. (1986) The taxonomy of *Enterobacteriaceae*, isolation of *Enter-*

obacteriaceae and preliminary identification. The genus *Salmonella*. In: *Identification of Enterobacteriaceae*, 4th edn (eds Edwards, P. & Ewing, W. H.) pp. 1–91, 181–318. Elsevier, New York.

FAO/WHO (1995) *Application of risk analysis to food standards issues*. Report of the Joint FAO/WHO Expert Consultation. World Health Organisation, Geneva. WHO/FNU/FOS/95.3.

FAO/WHO (1997a) *Risk management and food safety*. FAO Food and Nutrition paper no. 65. Food and Agriculture Organisation, Rome.

FAO/WHO (1997b) Codex Alimentarius ALINORM 97/13A, Vol. 1B, General Requirements (Food Hygiene) Supplement. Pre-publication.

FAO/WHO (1998) *The application of risk communication to food standards and safety matters*. FAO Food and Nutrition paper no. 70. Food and Agriculture Organisation, Rome.

FAO/WHO (1999) Risk assessment of microbiological hazards in foods. Report of a Joint FAO/WHO Expert Consultation, Geneva, 15–19 March 1999. See Internet site www.who.int/fsf/mbriskassess/Consultation99/reporam.pdf

FAO/WHO (2000a). *Report of the Joint FAO/WHO expert consultation on risk assessment of microbiological hazards in foods*, Rome, 17–21 July 2000. See Internet site www.fao.org.ES/ESN/pagerisk/rskpage.htm

FAO/WHO (2000b) Activities on risk assessment of microbiological hazards in foods. Preliminary document: *WHO/FAO Guidelines on hazard characterization for pathogens in food and water*. See Internet site www.fao.org/WAICENT/FAOINFO/ECONOMIC/ESN/pagerisk/mra006.pdf

Farber, J. M. and Peterkin, P. I. (1991) *Listeria monocytogenes*: a food-borne pathogen. *Microbiol. Rev.* **55**, 476–511.

Farber, J. M., Ross, W. H. & Harwig, J. (1996) Health risk assessment of *Listeria monocytogenes* in Canada. *Int. J. Food Microbiol.* **30**, 145–156.

Fazil, A. M., Lammerding, A. & Ellis, A. (2000b) A quantitative risk assessment model for *Campylobacter jejuni* on chicken. See Internet site www.who.int/fsf/mbriskassess/studycourse/index.html

Fazil, A., Lammerding, A., Morales, R., Vicari, A. S. & Kasuga, F. (2000a) Hazard identification and hazard characterization of *Salmonella* in broilers and eggs. See Internet site www.fao.org/WAICENT/FAOINFO/ECONOMIC/ESN/pagerisk/mra003.pdf

FDA (1998) A proposed framework for evaluating and assuring the human safety of the microbial effects of antimicrobial new animal drugs intended for use in food-producing animals. See Internet site www.fda.gov/cvm/fda/mappgs/whjatsnew.html

FDA (1999) Structure and initial data survey for the risk assessment of the public health impact of foodborne *Listeria monocytogenes*. Preliminary information is at Internet site vm.cfsan.fda.gov/~dms/listrisk.html

FDA (2000a) Draft risk assessment on the human health impact of fluoroquinolone resistant *Campylobacter* associated with the consumption of chicken. Revised 9 Feb 2000. See Internet site www.fda.gov/cvm/fda/mappgs/ra/risk.htm

FDA (2000b) Draft risk assessment on the public health impact of *Vibrio parahaemolyticus* in raw molluscan shellfish. See Internet site www.cfsan.fda.gov/~dms/vprisk.html

FDA (2001) Draft assessment of the public health impact of foodborne *Listeria monocytogenes* among selected categories of ready-to-eat foods. Center for Food Safety and Applied Nutrition (FDA) and Food Safety Inspection Service

(FSIS), United States Department of Agriculture. See Internet site www.food-safety.gov/~dms/lmrisk.html

Ferguson, N. M., Donnelly, C. A., Ghani, A. C. & Anderson, R. M. (1999) Predicting the size of the epidemic of the new variant of Creutzfeldt–Jakob disease. *Br. Food J.* **101**, 86–98.

Fleet, G. H., Heiskanen, P., Reid, I. & Buckle, K. A. (2000) Foodborne viral illness – status in Australia. *Int. J. Food Microbiol.* **59**, 127–134.

Forsythe, S. J. (2000) *The Microbiology of Safe Food.* Blackwell Science, Oxford. See Internet site www.blackwell-science.com/Microsafefood; companion Internet site science.ntu.ac.uk/external/foodmicrobiol.htm

Forsythe, S. J., Dolby, J. M., Webster, A. D. B. & Cole, J. A., (1988) Nitrate- and nitrite-reducing bacteria in the achlorhydric stomach. *J. Med. Microbiol.* **25**, 253–259.

Forsythe, S. J. & Hayes, P. R. (1998) *Food hygiene, Microbiology and HACCP*, 3rd edn. Aspen, Gaithersburg, USA.

Frenzen, P. D., Riggs, T. L., Buzby, J. C., Breuer, T., Roberts, T., Voetsch, D., Reddy, S. & the FoodNet Working Group. (1999) *Salmonella* cost estimate updated using FoodNet data. *FoodReview* **22**, 10–15.

FSIS (Food Safety Inspection Service) (1998) *Salmonella* Enteritidis risk assessment. Shell eggs and egg products. See Internet site www.europa.eu.int/comm/dg24/health/sc/scv/out26_en.html.

Gale, P. (1998) Quantitative BSE risk assessment: relating exposures to risk. *Lett. Appl. Microbiol.* **27**, 239–242.

Gale, P. (2001) A review: developments in microbiological risk assessment for drinking water. *J. Appl. Microbiol.* **91**, 191–205.

Gale, P., Young, C. Stanfield, G & Oakes, D. (1998) A review: development of a risk assessment for BSE in the aquatic environment. *J. Appl. Microbiol.* **84**, 467–477.

GATT (1994) The application of the Uruguay round of multilateral trade negotiations: the legal texts. World Trade Organisation, Geneva. ISBN 92-870-1121-4.

Gerba, C. P., Rose, J. B. & Haas, C. N. (1996) Sensitive populations: who is at the greatest risk? *Int. J. Food Microbiol.* **30**, 87–99.

Gibson, A., Bratchell, N. & Roberts, T. (1988) Predicting microbial growth: growth responses of salmonellae in a laboratory medium as affected by pH, sodium chloride and storage temperature. *Int. J. Food Microbiol.* **6**, 155–178.

Gill, C. O. & Phillips, D. M. (1985) The effect of media composition on the relationship between temperature and growth rate of *Escherichia coli. Food Microbiol.* **2**, 285.

Golnazarian, C. A., Donnelly, C. W., Pintauro, S. J., *et al.* (1989) Comparison of infectious dose of *Listeria monocytogenes* F5817 as determined for normal versus compromised C57B1/6J mice. *J. Food Protect.* **52**, 696–701.

Granum, A. F. & Lund, B. M. (1997) *Bacillus cereus* and its food poisoning toxins. *FEMS Microbiol. Lett.* **157**, 223–228.

Haas, C. N. (1983) Estimation of the risk due to low doses of microorganisms: a comparison of alternative methodologies. *Am. J. Epidemiol.* **118**, 573–582.

Haas, C. (1999) On modeling correlated random variables in risk assessment. *Risk Anal.* **19**, 1205–1213.

Haas, C. N., Rose, J. B. & Gerba, C. P. (1999) *Quantitative Microbial Risk Assessment.* John Wiley & Sons, New York.

Haas, C. N., Thayyar-Madabusi, A., Rose, J. B. & Gerba, C. P. (2000) Development of a dose-response relationship for *Escherichia coli* O157:H7. *Int. J. Food Microbiol.* **56**, 153–159.

Hald, T., Vose, D. & Wegener, H. C. (2001) Quantifying the contribution of animal-food sources to human salmonellosis in Denmark in 1999. See Internet site www.lst.min.dk/publikationer/publikationer/publikationer/campuk/camenga_ref.doc

Harrigan, W. F. & Park, R. A. (1991) *Making safe food. A Management Guide for Microbiological Quality.* Academic Press, London.

Hart, C. A. & Cunliffe, N. A. (1999) Viral gastroenteritis. *Curr. Opin. Infect. Dis.* **12**, 447–457.

Hauschild, A. H. W., Hilsheimer, R., Jarvis, G. & Raymond, D. P. (1982) Contribution of nitrite to the control of *Clostridium botulinum* in liver sausage. *J. Food Protect.* **45**, 500–506.

Heitzier, A., Kohler, H. E., Reichert, P. & Hamer, G. (1991) Utility of phenomenological models for describing temperature dependence of bacterial growth. *Appl. Environ. Microbiol.* **57**, 2656.

Hendrickx, M., Ludikhuyze, L., Vanden Broeck, I. & Weemaes, C. (1998) Effects of high pressure on enzymes related to food quality. *Trends Food Sci. Technol.* **9**, 197–203.

Holcomb, D. L., Smith, M. A., Ware, G. O., Hung, Y. C., Brackett, R. E. & Doyle, M. P. (1999) Comparison of six dose–response models for use with food-borne pathogens. *Risk Anal.* **19**, 1091–1100.

Hoornstra, E. & Notermans, S. (2001) Quantitative microbiological risk assessment. *Int. J. Food Microbiol.* **66**, 21–29.

Huisman, G. W. & Kolter, R. (1994) Sensing starvation: a homoserine lactone-dependent signalling pathway in *Escherichia coli. Science* **265**, 537–539.

ICMSF (International Commission on Microbiological Specifications for Foods) (1986) *Microorganisms in Foods*, Vol. 2. *Sampling for Microbiological Analysis: Principles and Specific Applications.* University of Toronto Press, Toronto.

ICMSF (1988). *Microorganisms in Foods*, Vol. 4. *HACCP in Microbiological Safety and Quality.* Blackwell, Oxford, England.

ICMSF (1996a) *Microorganisms in Foods*, Vol. 5. *Characteristics of Microbial Pathogens (Microbiological Specifications of Food Pathogens).* Blackie, London.

ICMSF (1996b) The International Commission on Microbiological Specifications for Foods: update. *Food Control* **7**, 99–101.

ICMSF (1997) Establishment of microbiological safety criteria for foods in international trade. *World Health Stat. Quart.* **50**, 119–123.

ICMSF (1998a) *Microorganisms in Foods*, Vol. 6. *Microbial Ecology of Food Commodities.* Blackie, London.

ICMSF (1998b) Principles for the establishment of microbiological food safety objectives and related control measures. *Food Control* **9**, 379–384.

ICMSF (1998c) Potential application of risk assessment techniques to microbiological issues related to international trade in food and food products. *J. Food Protect.* **61**, 1075–1086.

ILSI (1996) A conceptual framework to assess the risks of human disease following exposure to pathogens. ILSI Risk Science Institute Pathogen Risk Assessment working group report. *Risk Anal.* **16**, 841–848.

ILSI (1998a) *Food Safety Management Tools.* (eds J. L. Jouve, M. F. Stringer & A. C. Baird-Parker). Report prepared under the responsibility of ILSI Europe Risk

Analysis in Microbiology task force. International Life Sciences Institute, Belgium.

ILSI (1998b) Principles for the development of risk assessment of microbiological hazards under directive 93/43/EEC concerning the hygiene of foodstuffs. European Commission. Sept. 1997. Download from http://europa.eu.int.

ILSI (2000) Revised framework for microbial risk assessment. An ILSI Risk Science Institute workshop report. International Life Sciences Institute. See Internet site www.ilsi.org/file/mrabook.pdf

Inouye, S., Yamashita, K., Yamadera, S., *et al.* (2000) Surveillance of viral gastroenteritis in Japan: pediatric cases and outbreak incidents. *J. Infect. Dis.* **181**, S270–S274.

JEMRA (2000) Preliminary document: WHO/FAO guidelines on hazard characterization for pathogens in food and water. See Internet site www.fao.org/WAICENT/FAOINFO/ECONOMIC/ESN/pagerisk/mra006.pdf

Jones, J. E. (1993) A real-time database/models base/expert system in predictive microbiology. *J. Ind. Microbiol.* **12**, 268.

Jouve, J. L. (2001) Reducing the microbiological food safety risk: a major challenge for the 21st century. WHO Strategic Planning Meeting, Geneva, 20–21 February 2001. See Internet site www.who.int/fsf/mbriskassess/index.htm

Kalchayanand, N., Sikes, A., Dunne, C. P. & Ray, B. (1998) Factors influencing death and injury of foodborne pathogens by hydrostatic pressure-pasteurization. *Food Microbiol.* **15**, 207–214.

Kelly, L., Anderson, W. & Snary, E. (2000) Preliminary report: exposure assessment of *Salmonella* spp. in broilers. JEMRA. See Internet site www.fao.org/WAICENT/FAOINFO/ECONOMIC/ESN/pagerisk/mra005.pdf

Klapwijk, P. M., Jouve, J.-L. & Stringer, M. F. (2000) Microbiological risk assessment in Europe: the next decade. *Int. J. Food Microbiol.* **58**, 223–230.

Kleerebezem, M., Quadri, L. E. N., Kuipers, O. P. & De Vos, W. M. (1997) Quorum sensing by peptide pheromones and two-component signal-transduction systems in gram-positive bacteria. *Mol. Microbiol.* **24**, 895–904.

Knorr, D. (1993) Effects of high-hydrostatic pressure processes on food safety and quality. *Food Technol.* **47**, 156–162.

Kothary, M. H. & Babu, U. S. (2001) Infective dose of foodborne pathogens in volunteers: a review. *J. Food Safety* **21**, 49–73.

Kramer, J. M. & Gilbert, R. J. (1989) *Bacillus cereus* and other *Bacillus* species. In *Foodborne Bacterial Pathogens*, ed. M. P. Doyle, pp. 21–70. Marcel Dekker, New York.

Kwon, Y. M. & Ricke, S. C. (1998) Induction of acid resistance of *Salmonella typhimurium* by exposure to short-chain fatty acids. *Appl. Env. Microbiol.* **64**, 3458–3463.

Lammerding, A. M. (1997) An overview of microbial food safety risk assessment. *J. Food Protect.* **60**, 1420–1425.

Lammerding, A. M. & Fazil, A. (2000) Hazard identification and exposure assessment for microbial food safety risk assessment. *Int. J. Food Microbiol.* **58**, 147–158.

Lammerding, A. M. & Paoli, G. M. (1997) Quantitative risk assessment: an emerging tool for emerging foodborne pathogens. *Emerg. Infect. Dis.* **3**, 483–487.

Langeveld, L. P. M., van Spoosen, W. A., van Beresteijn, E. C. H. & Notermans, S. (1996) Consumption by healthy adults of pasteurised milk with a high concentration of *Bacillus cereus*: a double-blind study. *J. Food Protect.* **59**, 723–726.

Lederberg, J. (1997) Infectious disease as an evolutionary paradigm. *Emerg. Infect. Dis.* **3**, 417–423.

Li, S. Z., Marquardt, R. R. & Abramson, D. (2000) Immunochemical defection of molds: a review. *J. Food Protect.* **63**, 281–291.

Lindqvist, R. & Westöö, A. (2000) Quantitative risk assessment for *Listeria monocytogenes* in smoked or gravad salmon and rainbow trout in Sweden. *Int. J. Food Microbiol.* **58**, 181–196.

Lindroth, S. E. & Genigeorgis, C. A. (1986) Probability of growth and toxin production by nonproteolytic *Clostridium botulinum* in rockfish stored under modified atmospheres. *Int. J. Food Microbiol.* **3**, 167–181.

Lindsay, J. A. (1997) Chronic sequelae of foodborne disease. *Emerg. Infect. Dis.* **3**, 443–452.

Lund, B. M., Graham, A. F., George, S. M. & Brown, D. (1990) The combined effect of inoculation temperature, pH and sorbic acid on the probability of growth of non-proteolytic type B *Clostridium botulinum. J. Appl. Bacteriol.* **69**, 481–492.

Lundgren, O. & Svensson, L. (2001) Pathogenesis of rotavirus diarrhea, *Microbes Infect.* **3**, 1145–1156.

McClure, P. J., Boogard, E., Kelly, T. M., Baranyi, J. & Roberts, T. A. (1993) A predictive model for the combined effects of pH, sodium chloride and temperature, on the growth of *Brochothrix thermosphacta. Int. J. Food Microbiol.* **19**, 161–178.

McCullough, N. B. & Eisele, C. W. (1951a) Experimental human salmonellosis. I. Pathogenicity of strains of *Salmonella meleagridis* and *Salmonella anatum* obtained from spray-dried whole egg. *J. Infect. Dis.* **88**, 278–289.

McCullough, N. B. & Eisele, C. W. (1951b) Experimental human salmonellosis. II. Pathogenicity of strains of *Salmonella newport*, *Salmonella derby* and *Salmonella bareilly* obtained from spray-dried whole egg. *J. Infect. Dis.* **89**, 209–213.

McCullough, N. B. & Eisele, C. W. (1951c) Experimental human salmonellosis. III. Pathogenicity of strains of *Salmonella pullorum* obtained from spray-dried whole egg. *J. Infect. Dis.* **89**, 259–266.

McDonald, K. & Sun, D.-W. (1999) Predictive food microbiology for the meat industry: a review. *Int. J. Food Microbiol.* **52**, 1–27.

McLauchlin, J. (1990a) Human listeriosis in Britain, 1967–85: a summary of 722 cases. 1. Listeriosis during pregnancy and in the newborn. *Epidemiol. Infect.* **104**, 181–189.

McLauchlin, J. (1990b) Distribution of serovars of *Listeria monocytogenes* isolated from different categories of patients with listeriosis. *Eur. J. Clin. Microbiol. Infect. Dis.* **9**, 201–203.

McMeekin, T. A., Olley, J. N., Ross, T. & Ratkowsky, D. A. (1993) *Predictive Microbiology*. John Wiley & Sons, Chichester.

McNab, W. B. (1998) Review: a general framework illustrating an approach to quantitative microbial food safety risk assessment. *J. Food Protect.* **61**, 1216–1228.

Mansfield, L. P. & Forsythe, S. J. (2001) Demonstration of the Rb_1 lipopolysaccharide core structure in *Salmonella* strains with the monoclonal antibody M105. *J. Med. Microbiol.* **50**, 339–344.

Mantle, P. (1999) Risk assessment and the importance of ochratoxins. Paper presented at *Risk assessment and management of microbial hazards associated with food and water*, Joint meeting of Society for General Microbiology and Biodeterioration Society, Sheffield, 11–12 January 1999.

Marks, H. M., Coleman, M. E., Lin, J. C.-T & Roberts, T. (1998) Topics in microbial risk assessment: dynamic flow tree process. *Risk Anal.* **18**, 309-328.

Martin, S. A., Wallsten, T. S. & Beaulieu, N. D. (1995) Assessing the risk of microbial pathogens: application of a judgment-encoding methodology. *J. Food Protect.* **58**, 289-295.

Mead, P. S., Slutsker, L., Dietz, V., McCaig, L. F., Bresee, J. S., Shapiro, C., Griffin, P. M. & Tauxe, R. V. (1999) Food-related illness and death in the United States. *Emerg. Infect. Dis.* **5**, 607-625.

Medema, G. J. & Schijven, J. F. (2001) Modelling the sewage discharge and dispersion of *Cryptosporidium* and *Giardia* in surface water. *Water Res.* **35**, 4307-4316.

Medema, G. J., Teunis, P. F. M., Havelaar, A. H. & Haas, C. N. (1996) Assessment of the dose-response relationship of *Campylobacter jejuni*. *Int. J. Food Microbiol.* **30**, 101-111.

Meng, J. & Genigeorgis, C. A. (1993) Model lag phase of nonproteolytic *Clostridium botulinum* toxigenesis in cooked turkey and chicken breast as affected by temperature, sodium lactate, sodium chloride and spore inoculum. *Int. J. Food Microbiol.* **19**, 109-122.

Miller, A. J., Whiting, R. C. & Smith, J. L. (1997) Use of risk assessment to reduce listeriosis incidence. *Food Technol.* **51**, 100-103.

Mintz, E. D., Cartter, M. L., Hadler, J. L., Wassell, J. T., Zingeser, J. A. & Tauxe, R. V. (1994) Dose-response effects in an outbreak of *Salmonella* Enteritidis. Epidemiol. Infect. **112**, 13-19.

Mitchell, R. T. (2000) *Practical Microbiological Risk Analysis*. Chandos, Oxford.

Morris, J. G., Jr. & Potter, M. (1997) Emergence of new pathogens as a function of changes in host susceptibility. *Emerg. Infect. Dis.* **3**, 435-441.

Mortimore, S. & Wallace, C. (1994) *HACCP - A Practical Approach*. Practical Approaches to Food Control and Food Quality Series No. 1. Chapman & Hall, London.

Moss, M. O. (1999) Risk assessment for aflatoxins in foodstuffs. Paper presented at *Risk assessment and management of microbial hazards associated with food and water*, Joint meeting of Society for General Microbiology and Biodeterioration Society, Sheffield, 11-12 January 1999.

Mossel, D. A. A. (1988) Impact of foodborne pathogens on today's world and prospects for management. *Int. J. Food Microbiol.* **7**, 205-209.

NACMCF (National Advisory Committee on Microbiological Criteria for Foods) (1992) Hazard analysis and critical control point system. *Int. J. Food Microbiol.* **16**, 1-23.

NACMCF (National Advisory Committee on Microbiological Criteria for Foods) (1998a) Principles of risk assessment for illness caused by foodborne biological agents. *J. Food Protect.* **16**, 1071-1074.

NACMCF (National Advisory Committee on Microbiological Criteria for Foods) (1998b) Hazard analysis and critical control point principles and application guidelines. *J. Food Protect.* **61**, 1246-1259.

National Academy of Sciences, USA (1983) *Elements of Risk Assessment and Risk Management*. National Academy Press, Washington, DC, USA.

Nauta, M. J. (2000) Separation of uncertainty and variability in quantitative microbial risk assessment models. *Int. J. Food Microbiol.* **57**, 9-18.

Norman, J. (1999) Risk assessment for patulin in apples. Paper presented at *Risk assessment and management of microbial hazards associated with food and water*, Joint meeting of Society for General Microbiology and Biodeterioration Society, Sheffield, 11–12 January 1999.

Notermans, S. & Batt, C.A. (1998) A risk assessment approach for food-borne *Bacillus cereus* and its toxins. *J. Appl. Microbiol.* (Suppl.) **84**, 51S–61S.

Notermans, S., Dufrenne, J., Teunis, P., Beumer, R., te Giffel, M. & Peeters Weem, P. (1997) A risk assessment study of *Bacillus cereus* present in pasteurised milk. *Food Microbiol.* **14**, 143–151.

Notermans, S., Dufrenne, J., Teunis, P. & Chackraborty, T. (1998a) Studies on the risk assessment of *Listeria monocytogenes. J. Food Protect.* **61**, 244–248.

Notermans, S. Gallhoff, G., Zweitering, M. H. & Mead, G. C. (1995) The HACCP concept: specification of criteria using quantitative risk assessment. *Food Microbiol.* **12**, 81–90.

Notermans, S & van der Giessen, A. (1993) Foodborne diseases in the 1980's and 1990's: the Dutch experience. *Food Contam.* **4**, 122–124.

Notermans, S., Hoornstra, E., Northolt, M. D. & Hofstra, H. (1999) How risk analysis can improve HACCP. *Food Sci. Technol. Today* **13**, 49–54.

Notermans, S. & Mead, G. C. (1996) Incorporation of elements of quantitative risk analysis in the HACCP system. *Int. J. Food Microbiol.* **30**, 157–173.

Notermans, S., Mead, G. C. & Jouve, J. L. (1996) Food products and consumer protection: a conceptual approach and glossary of terms. *Int. J. Food Microbiol.* **30**, 175–183.

Notermans, S., Nauta, M. J., Jansen, J., Jouve, J. L. & Mead, G. C. (1998b) A risk assessment approach to evaluating food safety based on product surveillance. *Food Control* **9**, 217–223.

Notermans, S. & Teunis, P. (1996) Quantitative risk analysis and the production of microbiologically safe food: an introduction. *Int. J. Food Microbiol.* **30**, 9–25.

NRC (National Research Council) (1993) *Risk Assessment in the Federal Government: Managing the Process.* National Academy Press, Washington DC.

NRC (National Research Council) (1994) *Science and Judgement in Risk Assessment.* National Academy Press, Washington DC.

NRC (National Research Council) (1998) *Ensuring Safe Food from Production to Consumption.* National Academy Press, Washington DC.

Oscar, T. P. (1998a) Growth kinetics of *Salmonella* isolates in a laboratory medium as affected by isolate and holding temperature. *J. Food Protect.* **61**, 964–968.

Oscar, T. P. (1998b) The development of a risk assessment model for use in the poultry industry. *J. Food Safety* **18**, 371–381.

Oscar, T. P. (1999a) Response surface models for effects of temperature, pH, and previous growth pH on growth kinetics of *Salmonella typhimurium* in brain-heart infusion broth. *J. Food Protect.* **62**, 106–111.

Oscar, T. P. (1999b) Response surface models for effects of temperature, pH, and previous temperature on lag-time and specific growth rate of *Salmonella typhimurium* on cooked chicken breast. *J. Food Protect.* **62**, 1111–1114.

Oscar, T. P. (1999c) Response surface models for effects of temperature and previous growth sodium chloride on growth kinetics of *Salmonella typhimurium* on cooked chicken breast. *J. Food Protect.* **62**, 1470–1474.

Paoli, G. (Unpublished) Health Canada risk assessment model for *Salmonella* Enteritidis. (Quoted by Fazil *et al.* 2000a.)

Peeler, J. T. & Bunning, V. K. (1994) Hazard assessment of *Listeria monocytogenes* in the processing of bovine milk. *J. Food Protect.* **57**, 689–697.

Peterson, W. L., MacKowiak, P. A., Barnett, C. C., Marling-Cason, M. & Haley, M. L. (1989) The human gastric bactericidal barrier: mechanisms of action, relative antibacterial activity, and dietary influences. *J. Infect. Dis.* **159**, 979–983.

Ponka, A., Maunula, L., von Bonsdorff, C. H. & Lyytikainen, O. (1999) An outbreak of calicivirus associated with consumption of frozen raspberries. *Epidemiol. Infect.* **123**, 469–474.

Potter, M. E. (1996) Risk assessment terms and definitions. *J. Food Protect.* (Suppl.) 6–9.

Pruitt, K. M. & Kamau, D. N. (1993) Mathematical models of bacterial growth, inhibition and death under combined stress conditions. *J. Ind. Microbiol.* **12**, 221.

Rees, C. E. D., Dodd, C. E. R., Gibson, P. T., Booth, I. R. & Stewart, G. S. A. B. (1995) The significance of bacteria in stationary phase to food microbiology. *Int. J. Food Microbiol.* **28**, 263–275.

Richards, G. P. (2001) Enteric virus contamination of foods through industrial practices: a primer on intervention strategies. *J. Ind. Microbiol. Biotechnol.* **27**, 117–125.

Roberts, J. A. (1996) *Economic Evaluation of Surveillance*. Department of Public Health and Policy, London.

Roberts, T. A. & Gibson, A. M. (1986) Chemical methods for controlling *Clostridium botulinum* in processed meats. *Food Technol.* **40**, 163–171.

Robinson, A., Gibson, A. M. & Roberts, T. A. (1982) Factors controlling the growth of *Clostridium botulinum* types A and B in pasteurized, cured meat. V. Prediction of toxin production: non-linear effects of storage temperature and salt concentration. *J. Food Protect.* **17**, 727–744.

Robinson, D. A. (1981) Infective dose of *Campylobacter jejuni* in milk. *Br. Med. J.* **282**, 1584.

Rose, J. B., Haas, C. N. & Gerba, C. P. (1995) Linking microbiological criteria for foods with quantitative risk assessment. *J. Food Safety* **15**, 121–132.

Rose, J. B., Haas, C. N. & Regli, S. (1991) Risk assessment and control of waterborne giardiasis. *Am. J. Publ. Health* **81**, 709–713.

Rose, J. B. & Sobsey, M. D. (1993) Quantitative risk assessment for viral contamination of shellfish and coastal waters. *J. Food Protect.* **56**, 1043–1050.

Ross, T. & McMeekin, T. A. (1994) Review paper: predictive microbiology. *Int. J. Food Microbiol.* **23**, 241–264.

Ross, T., Todd, E. & Smith, M. (2000, Withdrawn) Preliminary report: exposure assessment of *Listeria monocytogenes* in ready-to-eat foods. JEMRA. See Section 4.4.2.

Ross, W. (Unpublished) From exposure to illness: building a dose–response model for risk assessment. (Quoted by Fazil *et al.* 2000a.)

Rowan, N. J., Anderson, J. G. and Smith, J. E. (1998) Potential infective and toxic microbiological hazards associated with the consumption of fermented foods. In: *Microbiology of Fermented Foods* (ed. B.J.B. Wood), pp. 263–307. Blackie Academic and Professional, London.

Ruthven, P. K. (2000) Food and health economics in the 21st century. Asia Pacific J. Clin. Nutr. **9** (Suppl.) S101–S102.

Schellekens, M., Martens, T. & Roberts, T. A. *et al.* (1994) Computer aided microbial safety design of food processes. *Int. J. Food Microbiol.* **24**, 1–9.

Schiff, G. M., Stefanovic, E., Young, E. C., Sander, D. S., Pennekamo, J. K. & Ward, R. L. (1984) Studies of Echovirus 12 in volunteers: determination of minimal infectious dose and the effect of previous infection on infectious dose. *J. Infect. Dis.* **150**, 858–866.

SCOOP (1998) *Reports on tasks for scientific co-operation. Microbiological Criteria: collation of scientific and methodological information with a view to the assessment of microbiological risk for certain foodstuffs.* Report of experts participating in Task 2.1, European Commission, EUR 17638. Office for Official Publications of the European Communities, Luxemburg.

Scudamore, K. A. Current views of the significance and risk of fumonisins in maize. Paper presented at *Risk assessment and management of microbial hazards associated with food and water*, Joint meeting of Society for General Microbiology and Biodeterioration Society, Sheffield, 11–12 January 1999.

Serra, J. A., Domenech, E., Escriche, I. & Martorelli, S. (1999) Risk assessment and critical control points from the production perspective. *Int. J. Food Microbiol.* **46**, 9–26.

Sethi, D., Cumberland, P., Hudson, M. J., *et al.* (2001) A study of infectious intestinal disease in England: risk factors associated with group A rotavirus in children. *Epidemiol. Infect.* **126**, 63–70.

Shaw, R. D. (2000) Viral infections of the gastrointestinal tract. *Curr. Opin. Gastroenterol.* **16**, 12–17.

Shier, W. T. (2000) The fumonisin paradox: a review of research on oral bioavailability of fumonisin B-1, a mycotoxin produced by *Fusarium moniliforme*. *J. Toxicol.-Toxin Rev.* **19**, 161–187.

Skirrow, M. B. (1991) Epidemiology of *Campylobacter* enteritis. *Int. J. Food Microbiol.* **12**, 9–16.

Slauch, J., Taylor, R. & Maloy, S. (1997) Survival in a cruel world: how *Vibrio cholerae* and *Salmonella* respond to an unwilling host. *Genes Develop.* **11**, 1761–1774.

Snyder, O. P., Jr (1995) HACCP-TQM for retail and food service operations. In: *Advances in Meat Research*, Vol. 10. *HACCP in Meat, Poultry and Fish Processing* (eds A. M. Pearson & T. R. Dutson), pp. 230–299. Blackie, London.

Soker, J. A., Eisenberg, J. N. & Olivier, A. W. (1999) Case study: human infection through drinking water exposure to human infectious rotavirus. ILSI Research Foundation. Risk Science Institute, Washington.

Sparks, P. & Shepherd, R. (1994) Public perceptions of the potential hazards associated with food production and food consumption: an empirical study. *Risk Anal.* **14**, 799–806.

Stringer, S. C., George, S. M. & Peck, M. W. (2000) Thermal inactivation of *Escherichia coli* O157:H7. *J. Appl. Microbiol.* **88**, (Suppl.) 79S–89S.

Surkiewicz, B. F., Johnson, R. W., Moran, A. B. & Krumm, G. W. (1969) A bacteriological survey of chicken eviscerating plants. *Food Technol.* **23**, 1066–1069.

Sutherland, J. P. & Bayliss, A. J. (1994) Predictive modeling of growth of *Yersinia enterocolitica*: the effects of temperature, pH and sodium chloride. *Int. J. Food Microbiol.* **21**, 197–215.

Sutherland, J. P., Bayliss, A. J. & Roberts, T. A. (1994) Predictive modeling of growth of *Staphylococcus aureus*: the effects of temperature, pH and sodium chloride. *Int. J. Food Microbiol.* **21**, 217–236.

Sutherland, J. P., Bayliss, A. J. & Roberts, T. A. (1995) Predictive modeling of growth of *Escherichia coli* O157:H7: the effects of temperature, pH and sodium chloride. *Int. J. Food Microbiol.* **25**, 29–49.

Tauxe, R. V. (1997) Emerging foodborne diseases: an evolving public health challenge. *Emerg. Infect. Dis.* **3**, 425–434.

te Giffel, M. C., Beumer, R. R., Granum, P. E. & Rombous, F. M. (1997) Isolation and characterisation of *Bacillus cereus* from pasteurised milk in household refrigerators in The Netherlands. *Int. J. Food Microbiol.* **34**, 307–318.

Teunis, P. F. M. & Havelaar, A. H. (1999) *Cryptosporidium* in drinking water: evaluation of the ILSI/RSI quantitative risk assessment framework. Report 284–530-006, National Institute of Public Health and the Environment, Bilthoven, The Netherlands.

Teunis, P., Havelaar, A., Vliegenthart, J. & Roessink, C. (1997) Risk assessment of *Campylobacter* species in shellfish: identifying the unknown. *Water Sci. Technol.* **35**, 29–34.

Teunis, P. F. M., van der Heijden, O. G., van der Giessen, J. W. B. & Havelaar, A. H. (1996) The dose–response relation in human volunteers for gastro-intestinal pathogens. Report 284-550-002, National Institute of Public Health and the Environment, Bilthoven, The Netherlands.

Teunis, P. F. M., Nagelkerke, N. J. D. & Haas, C. N. (1999) Dose response models for infectious gastroenteritis. *Risk Anal.* **19**, 1251–1260.

Thayer, D., Muller, W., Buchanan, R. & Philips, J. (1987) Effect of NaCl, pH, temperature, and atmosphere in growth of *Salmonella typhimurium* in glucose–mineral salts medium. *Appl. Environ. Microbiol.* **53**, 1311–1315.

Thomas, P. & Newby, M. (1999) Estimating the size of the outbreak of new-variant CJD. *Br. Food J.* **101**, 44–57.

Thomson, G. T. D., Derubeis, D., Hodge, M. A., *et al.* (1995) Post-*Salmonella* reactive arthritis: late clinical sequelae in a point-source cohort. *Am. J. Med.* **98**, 13–21.

Thorns, C. J. (2000) Bacterial food-borne zoonoses. *Rev. Sci. Tech. Off. Int. Epiz.* **19**, 226–239.

Todd, E. C. D. (1989) Preliminary estimates of costs of foodborne disease in Canada and costs to reduce salmonellosis. *J. Food Protect.* **52**, 586–594.

Todd, E. C. D. (1996) Risk assessment of use of cracked eggs in Canada. *Int. J. Food Microbiol.* **30**, 125–143.

Todd, E. C. D., Farber, J. M., Rivers, M.-A., Smith, M. & Ross, W. H. (1999) Quantitative risk assessment for *Listeria monocytogenes* in cabbage in Canada. Health Canada report. (Unpublished)

Todd, E. C. D. & Harwig, J. (1996) Microbial risk assessment of food in Canada. *J. Food Protect.* (Suppl.) S10–S18.

Turner, P. C., Nikiema, P. & Wild, C. P. (1999) Fumonisin contamination of food: progress in development of biomarkers to better assess human health risks. *Mutat. Res. Genet. Toxicol. Environ. Mutagen.* **443**, 81–93.

Van Gerwen, S. J. C. & Zwietering, M. H. (1998) Growth and inactivation models to be used in quantitative Risk Assessments. *J. Food Protect.* **61**, 1541–1549.

Van Schothorst, M. (1996) Sampling plans for *Listeria monocytogenes*. *Food Control* **7**, 203–208.

Van Schothorst, M. (1997) Practical approaches to risk assessment. *J. Food Protect.* **60**, 1439–1443.

Vose, D. (1996) *Quantitative Risk Analysis: A Guide to Monte Carlo Simulation Modelling.* John Wiley, New York.

Vose, D. (1997) The application of quantitative risk analysis to microbial food safety. *J. Food Protect.* **60**, 1416.

Vose, D. J. (1998) The application of quantitative risk assessment to microbial food safety. *J. Food Protect.* **61**, 640-648.

Voysey, P. A. (1999) Aspects of microbiological risk assessment. *New Food* **2**, 3-13.

Voysey, P. A. & Brown, M. (2000) Microbiological risk assessment: a new approach to food safety control. *Int. J. Food Microbiol.* **58**, 173-180.

Ward, R. L., Berstein, D. I. & Young, E. C. (1986) Human rotavirus studies in volunteers of infectious dose and serological response to infection. *J. Infect. Dis.* **154**, 871-877.

Webley, D. J., Jackson, K. L. & Mullins, J. D. (1997) Mycotoxins in food: a review of recent analyses. *Food Aust.* **49**, 439.

Wheller, J. G., Sethi, D., Cowden, J. M. *et al.* (1999) Study of infectious intestinal disease in England: rates in the community, presenting to general practice, and reported to national surveillance. *Br. Med. J.* **318**, 1046-1050.

Whiting, R. C. (1995) Microbial modeling in foods. *Crit. Rev. Food Sci. Nut.* **35**, 467-494.

Whiting, R. C. (1997) Microbial database building: what have we learned? *Food Technol.* **51**, 82-86.

Whiting, R. C. & Buchanan, R. L. (1997) Development of a quantitative risk assessment model for *Salmonella* Enteritidis in pasteurized liquid eggs. *Int. J. Food Microbiol.* **36**, 111-125.

WHO (1995) *Report of the WHO Consultation on Selected Emerging Foodborne Diseases*, Berlin, 20-24 March 1995. WHO/CDS/VPH/95.142. World Health Organisation, Geneva.

WHO/FAO (2000a) Preliminary document: WHO/FAO guidelines on hazard characterization for pathogens in food and water. Rijksinstituut voor volksgezondheid en milieu (RIVM), Bilthoven, The Netherlands. See Internet site www.fao.org/ES/ESN/pagerisk/riskpage.htm

WHO/FAO (2000b) The interaction between assessors and managers of microbiological hazards in foods. Kiel, Germany. WHO/SDE/PHE/FOS/007. See Internet site www.fao.org/ES/ESN/pagerisk/riskpage.htm

Wijtzes, T., van't Riet, K., in't Veld, K., Huis, J. H. J. & Zwietering, M. H. (1998) A decision support system for the prediction of microbial food safety and food quality. *Int. J. Food Microbiol.* **42**, 79-90.

Wimptheimar, L., Altman, N. S. & Hotchkiss, J. H. (1990) Growth of *Listeria monocytogenes* Scott A, serotype 4 and competitive spoilage organisms in raw chicken packaged under modified atmospheres and in air. *Int. J. Food Microbiol.* **11**, 205.

De Wit, M. A. S., Koopmans, M. P. G., Kortbeek, L. M., van Leeuwen, N. J., Bartelds, A. I. M. & van Duynhoven, T. H. P. (2001) Gastroenteritis in sentinel general practices, The Netherlands. *Emerg. Infect. Dis.* **7**, 82-91.

Yeh, F. S., Yu, M. C., Mo, C. C., Luo, S., Tong, M. J. & Henderson, B. E. (1989) Hepatitis B virus, aflatoxins, and hepatocellular carcinoma in Southern Guangxi, China. *Cancer Res.* **49**, 2506-2509.

Zink, D. L. (1997) The impact of consumer demands and trends on food processing. *Emerg. Infect. Dis.* **3**, 467-469.

Zwietering, M. H., de Wit, J. C. & Notermans, S. (1996) Application of predictive microbiology to estimate the number of *Bacillus cereus* in pasteurised milk at the point of consumption. *Int. J. Food Microbiol.* **30**, 55-70.

Internet Directory

Author's homepages
The Microbiology of Safe Food (Forsythe 2000) — www.theagarplate.com
The Microbial Risk Assessment companion site — www.blackwell-science.com/Microsafefood
@RISK (Monte Carlo simulation software) — science.ntu.ac.uk/external/foodmicrobiol.htm
www.palisade.com
Agreement of the Application of Sanitary and Phytosanitary Measures — www.wto.org/english/docs_e/legal_e/15-sps.pdf

Analytica (Monte Carlo simulation software) — www.lumina.com
CAST (Council for Agricultural Science and Technology) — www.cast-science.org
Codex Alimentarius Commission (CAC) — www.codexalimentarius.net/
CAC Principles of microbiological risk analysis — www.who.int/fsf/mbriskassess/pdf/draftpr.pdf
Codex Committee on General Principles, 15th Session, Paris, April 2000

Risk Analysis — europa.eu.int/comm/food/fs/ifsi/eupositions/ccgp/ccgp_item3a_en.html

Crystal Ball (Monte Carlo simulation software) — www.decisioneering.com/crystal_ball/index.html
Denmark *C. jejuni* in broilers — www.lst.min.dk/publikationer/publikationer/publikationer/campuk/cameng_ref.doc
europa.eu.int/comm/dg24/health/sc/scv/out25-en.pdf

European Commission
(1999a) *L. monocytogenes* — europa.eu.int/comm/dg24/health/sc/oldcomm7/out07_en.html
(1999b) Risk assessment and 93/43/EEC — europa.eu.int/comm/food/fs/sc/ssc/out82_en.html
Harmonization of risk assessment

FAO
JECFA — www.fao.org/es/ESN/Jecfa/Jecfa.htm
JEMRA — www.fao.org/WAICENT/FAOINFO/ECONOMIC/ESN/pagerisk/riskpage.htm

FAO/WHO documents
Risk Analysis, Geneva 1995 www.who.int/fsf/mbriskassess/applicara/index.htm
Risk Management & Food Safety, Rome 1997 www.fao.org/WAICENT/faoinfo/economic/esn/risk/
riskcont.htm
Risk Communication, Rome 1998 www.fao.org/waicent/faoinfo/economic/esn/riskcomm/
HTTOC.htm
Risk Assessment, Geneva 1999 www.who.int/fsf/mbriskassess/Consultation99/reporam.pdf
Risk Management 2000 www.fao.org/ES/ESN/pagerisk/riskpage.htm
General – Food Control www.fao.org/waicent/faoinfo/economic/esn/Control.htm
L. monocytogenes (Buchanan & Lindqvist 2000) www.fao.org/WAICENT/FAOINFO/ECONOMIC/ESN/pagerisk/
mra001.pdf
L. monocytogenes RTE (Ross *et al.* 2000) Withdrawn. See Section 4.4.2
Hazard identification and hazard characterization of www.fao.org/WAICENT/FAOINFO/ECONOMIC/ESN/pagerisk/
Salmonella in broilers and eggs (Fazil *et al.* 2000a) mra003.pdf
Exposure assessment of S. Enteritidis in eggs (Ebel *et al.* 2000) www.fao.org/WAICENT/FAOINFO/ECONOMIC/ESN/pagerisk/
mra004.pdf
Exposure assessment of *Salmonella* in broilers (Kelly *et al.* www.fao.org/WAICENT/FAOINFO/ECONOMIC/ESN/pagerisk/
2000) mra005.pdf
Hazard characterization for pathogens in food and water www.fao.org/ES/ESN/pagerisk/riskpage.htm.
Assesors & managers of microbiological hazards www.fao.org/ES/ESN/pagerisk/riskpage.htm
FDA Bad Bug Book vm.cfsan.fda.gov/~mow/intro.html
Fluoroquinolone resistant *Campylobacter* www.fda.gov/cvm/antimicrobial/ra/risk.html
L. monocytogenes risk assessment survey www.foodsafety.gov/~dms/lmrisk.html
V. parahaemolyticus in shellfish model www.cfsan.fda.gov/~dms/vprisk.html
Food Safety Risk Analysis Clearinghouse www.foodriskclearinghouse.umd.edu
Food Micromodel (predictive microbiology software) www.lfra.co.uk/lfra/micromod.html
FSIS
Salmonella in shell eggs risk assessment www.fsis.usda.gov/ophs/risk/contents.htm
S. Enteritidis risk assessment model www.fsis.usda.gov/ophs/risk/semodel.htm
E. coli O157 in beef risk assessment www.fsis.usda.gov/OPHS/ecolrisk/prelim.htm

Global *Salmonella* surveillance (Global Salm-Surv)	www.who.int/emc/diseases/zoo/SALM-SURV/SlideShow
International Life Sciences Institute (ILSI)	www.ilsi.org
ILSI revised framework for microbial risk assessment	www.ilsi.org/file/mrabook.pdf
MicroFit (predictive growth model)	www.ifrn.bbsrc.ac.uk/Microfit
NACMCF	seafood.ucdavis.edu/Guidelines/nacmcf.htm
Online training courses and model explanations	
WHO	www.who.int/fsf/mbriskassess/studycourse/index.html
Global perspective of risk assessment (Hogue)	
C. jejuni in chicken (Fazil *et al.*)	
Risk assessment advice (Schlundt)	
Monte Carlo simulations (Yoe)	
S. Enteritidis in egg and egg products (Wachsmuth)	
Quantitative risk modelling (Lammerding)	
FSIS *E. coli* O157:H7 risk assessment in ground beef (Powell)	www.fsis.usda.gov/ophs/ecolrisk/pubmeet/index.htm
Poultry FARM (Oscar)	www.arserrc.gov/mfs/PF2Instr.htm
FAO/WHO initiative on microbial risk assessment	www.who.int/fsf/mbriskassess/IAFP_meeting_01/index.htm
Overview of FAO/WHO initiative (Schlundt)	
Exposure assessment of *Salmonella* spp. in broilers (Kelly)	
Exposure assessment of *Salmonella* Enteritidis in eggs (Kasuga)	
Hazard and risk characterisation of *Salmonella* (Fazil)	
Exposure assessment of *L. monocytogenes* in ready-to-eat meat and fish (Ross)	
Exposure assessment of *L. monocytogenes* in dairy products (Todd)	
Hazard and risk characterisation of *L. monocytogenes* (Buchanan)	

Pathogen Modeling Program (predictive growth model)	www.arserrc.gov/mfs/pathogen.htm
Risk assessment frameworks	www.gov.on.ca/omafra/english/research/risk/frameworks/index.html
Risk assessment tools	www.foodriskclearinghouse.umd.edu/tools.htm
RiskWorld	www.riskworld.com
Society for Risk Analysis	www.sra.org/index.htm
USA Elimination of S. Enteritidis illness due to eggs	www.foodsafety.gov/~fsg/ceggs.html
USDA risk analysis bibliography	www.nal.usda.gov/fnic/foodborne/risk.htm
WHO antimicrobial resistance	www.who.int/emc/diseases/zoo/antimicrobial.html
WHO index	www.who.int/fsf/index.htm
WHO microbiological risk assessment	www.who.int/fsf/mbriskassess/index.htm

INDEX

@RISK, 103, 117, 129, 133, 138, 200
abortion, 8, 14
acceptable risk, 107, 175
achlorhydria, 99
acid tolerance, 53, 82, 86
adaptation, 11, 12, 53-5, 85, 86, 176
additives, 14, 16, 26, 46-8
Aeromonas hydrophila, 1, 3, 14
aflatoxins, 5, 19, 37, 38, 170, 171
Africa, 15, 170-72
age, 6, 12, 14, 15, 18, 34, 82, 84-6, 95,
     97, 99, 115, 123, 125, 136, 143,
     149, 151, 156, 159, 171, 175
AIDS, 14, 15, 154
algal toxins, 3
amoebiasis, 7
anacrobiosis, 54
animal feeds, 21
animal models, 71, 87, 88, 97, 101, 145,
     167
Anisakis simplexi, 11, 50
anorexia, 9
antibiotic, 10, 12, 21, 38, 40, 42, 84, 85,
     114, 115, 131, 203
   antibiotic resistant, 12, 84, 85,
     138-42
antigenic profile, 9, 85
apple juice, 156
appropriate level of sanitary or
     phytosanitary protection (ALOP),
     107
Arcobacter butzleri, 11
Argentina, 157
artherosclerosis, 9
arthritis, 2, 8, 9, 115, 118, 119, 123
as low as reasonably achievable
     (ALARA), 106, 107

Asia, 15, 171, 172
Aspergillus, 168, 170
   Asp. flavus, 169, 170
   Asp. oryzae, 171
   Asp. ochraceus, 169
   Asp. parasiticus, 169, 170
   Asp. sojae, 171
   Asp. tamari, 171
asymptomatic, 86, 95, 143, 173
atmosphere, 47, 57, 82
Australia, 18, 114, 157
autoimmune disease, 9, 10, 19, 115,
     132

Bacillus, 4, 32, 49, 52-4, 59, 161
   B. cereus, 2-5, 30, 32, 48, 56, 88, 89,
     93, 161-5, 178
   B. licheniformis, 163
   B. subtilis, 55, 163
bacteraemia, 2
baking, 49
Baranyi, 57, 58
barley, 4, 169, 172
beef, 2, 16, 17, 30, 136, 157, 161, 178,
     180, 201
Belgium, 14
benzoic acid, 47
beta-Poisson, 89-92, 94, 95, 101,
     120-23, 130, 159, 166, 174
biofilm, 50
biotechnology, 16
birds, 131 171
broilers, 73, 98, 109, 120, 121, 124,
     135, 136, 138, 141, 149, 177,
     200-202
Brucella abortus, 5, 45
brucellosis, 7, 8

bovine spongiform encephalitis (BSE), 11, 12, 14, 16, 19, 20, 24, 32, 112

Calicivirus, 6
Campylobacter, 4, 6, 9, 15, 18, 19, 32, 48, 96, 130-39, 141, 142,
  C. coli, 5, 130, 131, 136
  C. jejuni, 1-5, 9, 11, 12, 18, 48, 73, 87-91, 95, 96, 102, 109, 115, 124, 129-39, 142, 157, 176, 178, 180, 200, 202
  C. jejuni O19, 5, 132
  C. lari, 131
  C. upsaliensis, 131, 136
campylobacteriosis, 7, 8, 15, 19, 48, 136
Canada, 18, 97, 115, 121-3
cancer, 15, 154, 170-72
canning, 46, 54
carcinogens, 10, 17, 132, 169, 171, 172
cats, 113, 130, 136
cattle, 6, 15, 113, 130, 131, 136, 155, 156, 161, 170
cereals, 4, 169, 172
cheese, 3, 97, 143, 147-9, 151, 156, 163, 171, 178
chemical risk assessment, 69, 71, 72, 77, 108
chicken, 2, 30, 83, 124-7, 129, 132-4, 138, 139, 141, 142, 163, 178, 202
children (See also new-born, infants), 6, 14, 19, 115, 123, 154-6, 159, 160, 173
Chile, 157
chilled foods, 52, 54, 131
China, 172, 173
cholera, 7, 11, 19
chronic sequelae, 5, 8, 9, 19, 86, 87, 123, 136
cider, 53, 156
ciprofloxacin, 131, 138
Citrobacter fruendi, 9
clams, 174
Clonorchis, 48, 50
Clostridium, 4, 49, 52-4, 178
  Cl. botulinum, 2-5, 48, 52, 56, 89
  Cl. perfringens, 2-5, 18, 48, 52, 89, 93, 162
Codex Alimentarius Commission (CAC), 19, 22-31, 36, 37, 45, 47, 68, 69, 72, 75, 105, 107, 113, 125, 144, 145, 149, 172, 174, 181, 200

Codex Committee on Food Hygiene (CCFH), 25-8, 68, 72, 107, 180
cold adaptation, 54
cold shock proteins (CSP), 53-5
colitis, 8, 9, 155
consumer, 2, 13, 14, 16, 20, 22, 23, 25, 28-30, 34, 36, 39, 40, 43, 61, 64, 65, 74, 78, 80, 82, 103, 106, 107, 109-11, 120, 125, 127, 132, 133, 138, 144, 149, 151, 152, 157, 159-61, 176, 182, 183
consumer's risk, 61, 64, 65
consumption, 12-14, 18, 24, 33, 36, 37, 39, 52, 64, 66, 67, 71, 76, 78, 80, 82, 83, 87, 98, 109, 117, 118, 120, 125, 127-31, 136, 138, 141, 144, 148-50, 152, 156-8, 163-6, 174, 176, 181
consumption patterns, 12, 14, 78, 79, 80, 82, 83
contaminated food, 2, 6, 19, 98, 162, 171
cooking, 16, 42, 49, 51, 52, 80, 82, 92, 125, 127-9, 133, 153, 155, 157-9, 161, 166, 173, 176, 181
corrective action, 36, 43, 45
costs of food poisoning, 16, 18-21, 86
cream, 3, 116, 165
critical control points (CCP) 20, 24, 34-6, 40-45, 51, 60, 68, 69, 175
critical limits, 24, 36, 42-5, 51, 60, 71
Crohn's disease, 9, 10
cross-contamination, 81, 114, 116, 118, 124, 125, 130, 131, 135-7, 142, 155-7
Cryptosporidium parum, 3, 5, 11, 30, 89, 90, 94, 98, 178
Crystal Ball, 103, 138, 200
Cyclospora cayetanensis, 5, 11, 86, 178
Czech Republic, 157

D value, 48, 50, 51, 53, 56, 77, 98, 104, 118, 127, 128, 131, 133, 134, 153, 154, 162, 181, 184
dairy products, 3, 21, 53, 116, 151, 162, 178, 202
death kinetics, 50
decision trees, 40
Denmark, 109, 129, 135-7, 176, 200
detection methods, 18, 24, 94, 174

detergents, 50
deterministic approach, 102
detoxification, 49
developing countries, 6, 7, 14, 15, 19,
    24, 45, 176, 179
diarrhoea, 6, 9, 15, 19, 114, 131, 141,
    155-7, 162, 163, 165, 173
    diarrhoeal toxins, 93, 162
dioxin, 13, 14, 16, 19, 38, 45
DNA gyrase, 138
domestic animals, 68, 130, 142
dose-response, 29, 67, 71, 74, 75, 79, 81,
    83, 86-95, 97, 98, 100, 101, 108,
    121, 123, 125, 132, 133, 136, 137,
    145, 146, 148, 149, 157, 159, 163,
    164, 166, 167, 174-7, 181, 182
dried fruits, 172
drinking water, 30, 32, 136
drying, 55, 84, 143

echovirus, 15, 94, 174
economy, 6, 10, 14, 19, 20, 25, 82, 86,
    109, 200, 201
egg, 30, 32, 51, 62, 63, 73, 98, 116-21,
    123, 130, 135, 149, 177, 178, 180,
    201-203
elderly, 2, 6, 8, 14, 15, 29, 80, 81, 96,
    99, 119, 143, 145, 147, 148, 149,
    151, 154-6
emerging pathogens, 10, 11, 22, 24,
    138, 155
emetic toxins, 5, 93, 162, 163
end-product testing, 24, 35, 57, 61
England and Wales, 14, 18, 173
enrofloxacin, 131, 138
Enter-Net, 21
enteric infections, 9, 115, 140
Enterococcus, 138
enterocytes, 93, 144, 165
Enterobacter, 9, 11
    Ent. sakazakii, 5, 11
Enterobacteriaceae, 63, 114
epidemic, 10, 13, 19
epidemiology, 11, 66
ergotism, 170
Escherichia coli 1-12, 15, 16, 18, 20,
    30, 32, 48, 50-53, 55, 56, 59, 89,
    91-6, 115, 132, 154-61, 178, 180,
    202
    Diffusely adherent *E. coli* (DAEC),
    155

E. coli O157:H7, 1-3, 9, 11, 12, 15,
    16, 18, 20, 30, 32, 48, 50-53, 55,
    56, 59, 89, 91, 92, 95, 96, 132,
    154-61, 178, 201, 202
E. coli non-O157, 157
Enteroaggregative *E. coli* (EAEC), 11
Enterohaemorrhagic *E. coli* (EHEC),
    1, 5, 8, 154, 155, 178, 180
Enteroinvasive *E. coli* (EIEC), 1, 155
Enteropathogenic *E. coli* (EPEC), 1,
    5, 8, 155
Enterotoxigenic *E. coli* (ETEC), 1, 5,
    155
Shiga-toxigenic *E. coli* (STEC), 155
Verotoxigenic *E. coli* (VTEC), 155
ethylene bromide, 48
ethylene oxide, 49
Europe, 16, 18, 21, 29, 83, 155, 200
European Public Health and Food
    Safety Authority, 21
European Union, 29, 75, 151
exponential growth, 53, 57, 58,
exponential model, 89, 90, 92, 94,
    145-7
exposure assessment, 29, 69, 71,
    73-6, 78-83, 99-101, 123, 124,
    132, 145, 148-51, 166, 168,
    174-83, 202
extrinsic parameters, 47, 67, 80, 149

fair trade, 61, 74, 183
farm to fork, 21, 34, 66-8, 78, 79, 117,
    124, 132, 153, 157, 176
Food and Agriculture Organisation
    (FAO), 20-22, 24-8, 68, 69, 72, 73,
    83, 88, 98, 105, 177, 200-202
fat, 13, 50, 53, 54, 84-6, 144
Food and Drug Administration (FDA),
    30, 32, 40, 92, 103, 105, 138, 140,
    145, 146, 149, 156, 166-8, 201
foetus, 14, 143, 144, 149
fish, 3, 11, 45, 48, 50, 116, 143, 151,
    161-3, 165, 166, 171, 178, 202
flavour, 49
flow diagram, 36, 39, 40, 42, 78, 83
fluoroquinolone, 12, 32, 131, 138-42,
    201
food additives, 16, 26
food-borne, 1, 2, 4-10, 12-16, 18-22,
    24, 25, 27-9, 34, 37, 40, 44, 46, 47,
    49, 50, 54, 66-9, 71-3, 76-8, 83,

86, 87, 94, 95, 97, 117, 133, 136, 138, 149, 163, 165, 176, 203
food handlers, 14, 22, 34, 173
food handling, 14
Food Micromodel, 201
food matrix, 82-4, 87, 123, 144, 145
food processing, 11, 66, 78, 86, 105, 108, 110, 172
food quality, 55, 105
food safety objective, 30, 32, 61, 69, 70, 107, 109, 110, 151, 152, 177, 182
food scares, 16, 111
food vectors, 18
FoodNet, 18, 140
foreign travel, 117
formal report, 74, 75, 101
France, 16, 29, 155
freezing, 37, 39, 46, 50, 55, 136, 143, 167
fruit, 3, 19, 38, 53, 73, 172, 173
Food Safety and Inspection Service (FSIS), 30, 32, 51, 73, 117, 119-23, 161, 178, 201, 202
Fusarium, 168, 172
  F. culmorum, 172
  F. graminearium, 172
fumonisins, 11, 172

gastric acid, 15, 99
gastritis, 9
gastroenteritis, 2, 6, 7, 10, 15, 17-19, 48, 115, 116, 131, 132, 135, 137, 142, 146, 165, 173
General Agreeement on Tariffs and Trade (GAAT), 23, 26
gene, 11, 53, 115, 155, 165
genetic susceptibility, 9
genetically modified, 19, 20, 45, 112
genome, 11
Germany, 114, 157
Giardia lamblia, 3, 4, 10, 90, 94, 96
globalisation, 11-15, 19-22, 66, 177, 178
Gompertz equation, 57-9, 92, 93, 145, 147, 158, 159, 166
Good Hygienic Practice (GHP), 34, 35, 67, 69, 70, 84, 109
Good Manufacturing Practice (GMP), 24, 34, 35, 69, 70, 109, 110, 164
grain, 19
Graves disease, 9

ground beef, 2, 157, 161, 178, 180
guidelines, 22, 23, 25-8, 67, 108, 110, 175, 182, 183
Guillain-Barré syndrome (GBS), 2, 5, 8, 9, 19, 86, 132, 133

haemolysin, 165, 167
haemolytic anaemia, 9, 156
haemolytic uraemic syndrome (HUS), HUS, 5, 8, 9, 155-60
haemorrhagic colitis (HC), 155, 201
hamburger, 91, 92, 158, 160, 161
handling, 13, 14, 28, 33, 34, 45, 67, 68, 80, 105, 110, 116, 157, 166, 173, 176
Hazard Analysis Critical Control Point (HACCP), 20, 21, 24, 28, 32, 34-40, 42, 44, 45, 51, 60, 61, 64, 66-71, 74, 84, 108, 109, 153, 175
hazard characterisation, 27, 29, 71, 73, 75-8, 80, 81, 83, 84, 86, 98, 99, 113, 120, 121, 125, 144, 163, 166, 174, 179, 182, 183
hazard identification, 29, 71, 73-5, 77, 120, 121, 182
Health Canada, 121-3
heat shock proteins (HSP), 53-5 , 86
heat treatment, 4, 49, 52-4, 81, 104, 114, 127, 128, 152, 167
Helicobacter pylori, 9
helminths, 11, 48, 50, 182
hepatitis A, 1-3, 5, 10, 45, 47, 89
hepatitis B, 171, 172
hepatitis E, 11
high risk foods, 82
HLA-B27, 115
home, 12, 14, 82, 133, 160, 176
host factors, 83, 132
host susceptibility, 12, 15, 87
humidity, 47, 82, 170
hurdle technology, 53, 54
hygiene, 6, 12, 14, 23-5, 32, 68, 114, 130, 155

ice cream, 3, 97, 148
International Life Science Institute (ILSI), 29-31, 35, 69, 70, 72, 202
immune response, 2
immune status, 12, 14, 85-7, 95, 97, 99
immune system, 6, 9, 14, 15, 84, 87, 99, 143

immuno-compromised, 2, 6, 80, 81, 86, 146, 147, 152
immunodeficient, 28, 86
India, 16, 172
infants (*see also* new-borns, children), 6, 28, 80, 81, 119, 156, 173
infectious dose, 17, 88, 89, 93, 97, 115, 116, 121, 128, 129, 132, 156
infection threshold, 17, 87–92, 107, 108, 159, 183
infectious disease triangle, 83, 85
ingested dose, 12, 90, 92, 95, 99, 101, 123, 133, 158, 159, 163
ingredients, 37, 40, 44, 69, 112, 116
International Commission Microbial Specifications for Foods (ICMSF), 4, 5, 30, 36, 40, 47, 61, 62, 63, 77, 107
international trade, 12, 13, 20, 22, 23, 25, 30, 33, 36, 61, 66, 72, 109
intestinal tract, 4, 53, 93, 122, 145, 156
intrinsic parameters, 47, 54, 67, 80, 84
irradiation, 47, 48, 50, 181
ISO standards, 34, 35, 36
Italy, 157

Japan, 13, 16, 50, 114, 157, 165, 173
JECFA, 26, 172, 200
JECFI, 47
JEMRA, 26, 72, 73, 98, 113, 121, 122, 145, 146, 177, 200
Johne's disease, 10

Kaufmann-White, 114
kidney, 2, 155, 156, 170
Klebsiella, 9

labelling, 34
legumes, 4
life threatening, 5, 8, 21, 96, 116
lipopolysaccharide (LPS), 114
Listeria, 14, 16, 73, 142, 148
L. monocytogenes, 1–5, 10–12, 14, 18, 30, 32, 48, 52, 55, 59, 92, 96, 97, 129, 142–8, 150–54, 176, 178, 180, 200–202
listeriosis, 8, 15, 143, 145, 146, 151
livestock, 130, 135

mad cow disease, 16
maize, 169

malabsorption, 6, 9
malnutrition, 6, 15
mathematical models, 88, 145, 157, 177
meal size, 67, 149
mechanistic approach, 98–100
media, 17, 57
medical costs, 18–20
medical treatment, 15, 21, 114
metals, 13, 38
mice, 92, 148
microbial ecology, 80, 149
microbiological criteria, 4, 32, 61, 107, 110, 124, 177
microbiological risk assessment support system, 180
MicroFit, 58, 202
milk, 3, 10, 30, 40, 42, 45, 46, 88, 116, 130, 131, 136, 142–4, 146, 148, 151, 155, 156, 161–5, 169, 170
moisture, 43, 82
molecular mimicry, 10, 132
Monte Carlo, 101–104, 129, 130, 133, 151, 154, 159, 174, 178, 200, 202
morbidity, 87, 99, 100, 145, 154
mortality, 2, 5, 14, 15, 86, 87, 95, 98–100, 118, 144–6, 156, 158–60, 173, 177
moulds, 19, 45, 171, 182
mouse, 149, 169
Mycobacterium paratuberculosis, 9–11
mycotoxin, 3, 4, 11, 13, 38, 45, 168–72

NACMCF, 37, 40, 44, 69, 202
Norwalk-like viruses, 1, 3, 5, 11, 47, 89
NASA, 35
National Academy of Sciences (NAS), 31, 69
Netherlands, 7, 114, 135, 165
new-born babies (see also children, infants), 14, 15, 143, 145–9
nisin, 49
nitrite, 50, 59
nuts, 4, 170, 172

oats, 172
ochratoxin, 11, 38, 169, 170, 172
operating characteristic curve, 62–4, 177

orange juice, 53
osmotic stress, 55
outbreaks, 17, 19–22, 40, 49, 57, 71,
    73, 82, 97, 98, 111, 115, 121–3,
    131, 135, 142, 156, 157, 161, 163,
    166, 173, 176, 177
outbreak data, 87, 101, 146
overseas travel, 136
oysters, 167, 174

packaging, 20, 38, 39, 42, 47, 66, 82,
    129
parabens, 47
parasites, 37, 48, 50, 90, 173
pasta, 162, 163
pasteurisation, 4, 21, 40, 42, 47, 114,
    118, 130, 142, 155, 176
Pathogen Modeling Program, 59, 60,
    154, 203
pathogenesis, 144, 154
pathogenicity islands, 11
patulin, 38, 170, 171
Penicillium, 168, 170, 172
    P. camembertii, 171
    P. roqueforti, 171
personal hygiene, 6, 12, 114, 130, 155
personnel, 35, 37, 39, 40, 42, 177
Peru, 19
pesticide contamination, 13, 26, 38, 45
pH, 32, 37, 39, 42, 43, 47, 48, 53, 54,
    57–60, 77, 82, 85, 98, 99, 127, 131,
    159, 161, 162
phase variation, 9
placenta, 143
Poisson model, 89–92, 94, 95, 101, 120,
    121, 123, 130, 159, 166, 174
poliomyelitis, 7
polychlorinated biphenyls (PCB), 13,
    45
pork, 2, 48, 59, 130, 136, 172, 180
potato, 3, 162, 163
poultry, 2, 16, 19, 21, 32, 48, 73, 102,
    113, 116, 124, 128–31, 135, 136,
    138, 139, 143, 176, 178, 180
poultry FARM model, 32, 129, 132,
    178, 202
predictive microbiology, 24, 29, 55–60,
    69, 78, 79, 82, 101, 125, 126, 129,
    149, 154, 158, 159, 176, 203
pregnant women, 6, 7, 14, 15, 80, 81,
    85, 96, 119, 142–4, 154

preparation, 12, 14, 23, 34, 45, 49, 50,
    53, 66, 67, 69, 80, 82, 118, 125,
    138, 156, 176, 182
preservation methods, 12, 14, 46, 49,
    54, 55
preservatives, 16, 37, 38, 46, 47, 53, 54
pressure technology, 49
prions, 11
probabilistic model, 101, 102
probability of acceptance, 62, 64, 65
probability of illness, 71, 74, 79, 83, 92,
    120, 123, 133, 137, 159, 160, 177
probability of infection (Pi), 17, 79, 87,
    90–93, 95, 96, 101, 126, 127, 134,
    146
Probit, 166
Process Risk Model, 29, 132, 157
processed foods, 47, 54, 116, 131
processed meats, 131
producer's risk, 61, 64, 65
productivity losses, 18, 20
propionic acid, 47
proteinaceous, 86
protozoa, 38, 45, 48, 73, 89, 90, 182
Pseudoterranova decipens, 11
public awareness, 18
public health, 2, 10, 20, 21, 24, 25, 46,
    67, 76, 77, 108, 117–19, 121, 129,
    149, 166, 167, 172, 180

Quality assurance (QA) systems, 70,
    109
quantitative (microbial) risk
    assessment, 67, 73, 74, 78, 129,
    153, 157, 165, 174, 178, 182

raw ingredients, 13, 40, 45, 78, 80, 110,
    116, 153, 154, 182
reactive arthritis, 2, 9, 115, 118, 119,
    123
ready-to-eat foods (RTE), 73, 144, 145,
    149–52, 163–5, 178, 180, 202
regulatory authorities, 34, 36, 40, 52
Reiter's syndrome, 8, 9, 115
Reoviridae, 173
residues, 13, 26, 38, 45, 72
rice, 4, 162, 163
risk analysis, 4, 21, 24, 26–9, 60, 66,
    68–70, 74, 101, 103, 109–11,
    181–3, 200, 201, 203
risk assessor, 28, 74, 108, 177, 183

risk characterisation, 29, 31, 71, 74-6, 80, 81, 99–101, 132, 151, 166, 167, 174, 175, 177, 178, 182, 183, 202

risk communication, 27-9, 68, 70, 74, 110, 111, 152, 181-3, 201

risk estimate, 71, 75, 76, 77, 99–102, 149, 174, 182, 183

risk factor, 12, 105, 115, 136, 156, 160

risk management, 16, 24-30, 68, 70, 72, 74-6, 101, 103, 105 12, 152, 181-3, 201

risk manager, 25, 28, 69, 72, 74, 77, 83, 103, 105, 183

risk mitigation strategies, 67, 69, 76, 103, 120, 135, 138, 161, 166, 167

risk of death, 17, 107

risk perception, 22, 74, 110, 112, 183

rodents, 113, 131, 171

rotaviruses, 1, 6, 15, 30, 94, 96, 173, 174

Salmonella, 1-4, 6, 9, 15, 16, 18, 19, 21, 30, 32, 48, 56, 59, 63, 73, 86, 87, 89-91, 94-100, 104, 113-17, 120-31, 138, 142, 149, 176-8, 180, 201, 202
  S. Agona, 115
  S. Anatum, 122
  S. Bareilly, 122, 123
  S. Cholerasuis, 115
  S. Derby, 121, 122
  S. Dublin, 114
  S. Enteritidis, 5, 11, 30, 48, 51, 73, 104, 115-24, 130, 165, 178, 201-203
  S. Hadar, 115
  S. Heidelberg, 115
  S. Meleagridis, 121, 122
  S. Montevideo, 115
  S. Newport, 121, 122, 123
  S. Paratyphi, 5, 113, 116
  S. Typhimurium, 5, 9, 11, 12, 53, 55, 56, 59, 115, 123
salmonellosis, 7, 8, 15, 18, 48, 97, 114, 116, 121, 129
Salm-Surv, 21, 202
salt, 43, 53, 60
sampling plan, 4, 43, 61-4, 110
SCOOP, 29

seafood, 3, 48, 73, 143, 151, 165, 178

sentinel, 7, 18, 136, 176

sequelae, 2, 5, 8, 9, 19, 76, 86, 87, 95, 116, 123, 136, 177

shelf-life, 14, 46, 47, 152

shellfish, 2, 3, 37, 38, 105, 165-8, 174, 178, 180, 201

Shigella, 1, 3-5, 96, 178
  Sh. dysenteria, 92, 155
  Sh. flexneri, 60, 69, 89, 92, 115

shigellosis, 7, 8, 48

shrimps, 48

signal transduction, 53

small round structured viruses (SRSV; see Norwalk-like viruses), 6, 178

sorbic acid, 47

sorghum, 172

sous vide, 14

South-East Asia, 171, 172

spoilage, 2, 42, 46, 48-50

spore-forming bacteria, 4, 38, 42, 85, 89

spores, 4, 49-54, 143, 161, 163

Sanitary and Phytosanitary (SPS) Agreement, 23-5, 72, 107, 200

Staphylococcus aureus, 2-5, 18, 48, 52, 55, 56, 58, 89, 93, 163, 178

starvation, 53

statement of purpose, 74, 75, 77

statistics, 40, 55, 86, 97, 129, 146

sterilisation, 21, 142

stomach, 53, 82, 84-6, 97-9, 115, 144

storage, 13, 14, 34, 39, 42, 45, 55, 67, 69, 80, 85, 108, 110, 118, 124, 126-9, 143, 149-53, 158, 160, 161, 163, 167, 176, 183

storage tests, 82

Streptococcus, 10
  Strep. parasanguinis, 11

stress factors, 53

sulphites, 47

surveillance, 10, 22, 71, 77-9, 97, 101, 103, 119, 129, 141, 160, 165, 176, 202

susceptible populations, 2, 6, 12, 15, 82, 86, 119, 121, 153

sushi, 13

Sweden, 21

symptoms, 1, 2, 7, 88, 95, 96, 98, 99, 114-16, 131, 132, 154, 162, 163, 165

Taenia saginata, 5, 8, 48, 50
tapeworm, 2, 23
technical barriers to trade (TBT)
    agreement, 23, 25
temperature abuse, 52, 60, 84, 125,
    126, 160, 161, 166, 176
thermal death, 50
threshold values, 17, 87–92, 107, 108,
    154, 159, 183
thrombic thrombocytopenic purpura
    (TTP), 155
toilet, 6
tolerable risk, 107
tornado chart, 101, 103, 105, 135, 167
Total Quality Management (TQM), 34,
    35, 69, 70
toxins, 1–4, 10, 29, 34, 35, 37–9, 45, 46,
    50, 52, 56, 77, 82, 93, 103, 156,
    162, 170–72, 182
Toxoplasma gondii, 5, 11, 48
toxoplasmosis, 8, 9
training, 34, 70, 177, 178, 202
triangular distribution, 102, 133
Trichinella spiralis, 2, 5, 48, 50
trichothecenes, 11, 172
turkey, 2, 136
typhoid, 7, 14, 116, 123

UK, 7, 16, 58– 60, 112, 136, 155
uncertainty, 28, 74-6, 77, 87, 92, 99,
    100–103, 106, 112, 133, 135, 152,
    159, 180–84
United States of America (USA) 7, 15,
    18, 19, 30, 35, 48, 86, 97, 112, 114,
    117, 132, 155-7, 203
United States Department of
    Agriculture (USDA), 59, 120, 149,
    203
unpasteurised, 136, 148, 151
Uruguay Round Agreement, 13, 23, 72

variability of values, 8, 61, 62, 75, 81,
    83, 87, 92, 95, 100–103, 135, 145,
    151, 161, 184

variant CJD (vCJD), 11, 16, 19, 20, 24,
    112
vegetables, 2, 19, 48, 73, 142, 143, 148,
    151, 162, 163, 178
verotoxins, 156
Vibrio, 3, 48, 73, 165, 166, 178
    V. cholerae, 5, 11, 89, 95, 96
    V. parahaemolyticus, 5, 32, 48, 89,
        103, 105, 165–8, 178, 180
    V. vulnificus, 5, 11
virulence, 2, 11, 12, 53, 81, 84, 85, 87,
    93, 96, 99, 114, 131, 165, 167
viruses, 1–3, 5, 6, 10, 11, 15, 22, 30, 37,
    38, 45, 47, 73, 82, 89, 94, 96, 173,
    174, 178, 182
volunteer (human) feeding trials, 71,
    86, 88, 91, 92, 95–8, 101, 121, 122,
    133, 145, 163, 167, 174,

water activity (a_w), 32, 37, 39, 42, 43,
    47, 48, 55, 57–9, 77, 82, 85, 131,
    158, 159, 161, 162
water-borne, 4, 6, 10, 30, 73, 94–6
Weibull-gamma, 90, 92, 121, 146
wheat, 169, 172
World Health Organisation (WHO),
    19–22, 24–9, 37, 66, 68, 69, 72, 73,
    83, 86, 88, 98, 105, 177, 178,
    201–203
World Trade Organisation (WTO) 22,
    23, 25

yogurt, 156
Yersinia, 8, 9, 14
    Y. enterocolitica, 2, 3, 5, 10, 48, 52,
        56, 59, 89, 115

z value, 50, 52, 184
zero risk, 16
zero tolerance, 148, 149, 151, 164
zearolenone, 11, 169, 172